The R.A.M.S. Library of Alchemy

Volume 29

The Book of
Abraham the Jew

by

Abraham Eleazar

R.A.M.S. Publishing Company

The Book of
Abraham the Jew

by

Abraham Eleazar

Produced by

Restorers of Alchemical Manuscripts Society
1982

R.A.M.S. Publishing Company

R.A.M.S. Publishing Company
117 Rutherford Lane
Stuarts Draft VA 24477

The Book of Abraham the Jew
Copyright © 2015 R.A.M.S. Publishing Company

First Edition 2015

ISBN-13 **978-1511667166**
ISBN-10 **1511667168**

Image Processing by Philip N. Wheeler

Printed in the United States of America

Table of Contents

Dedicated to Hans W. Nintzel,

American Alchemist

and

Founder of the

Restorers of Alchemical Manuscripts Society

(R.A.M.S.)

Disclaimer

Liability: The publisher does not warrant or assume any legal liability or responsibility for the accuracy, completeness, or usefulness of any information, apparatus, product, or process disclosed. The publisher makes no representation as to the accuracy or completeness of the contents of this book and specifically disclaims any implied warranty of merchantability or fitness for a particular purpose. No warranty may be created or extended by written sales materials or sales representatives. You should obtain professional consultation where appropriate. The publisher shall not be liable for any loss of profit or other commercial or personal damages, including but not limited to special, incidental, consequential, or other damages.

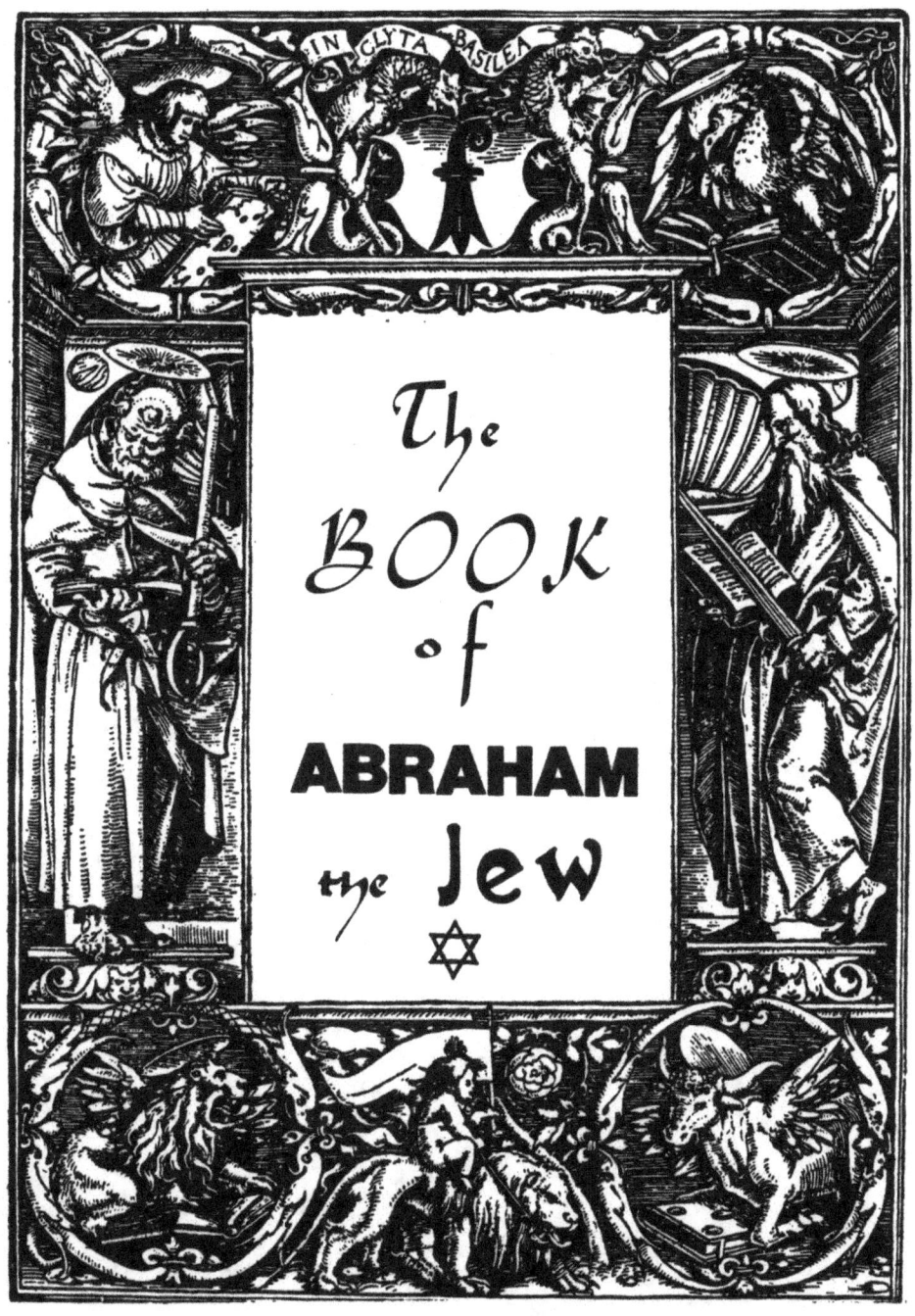

The BOOK of ABRAHAM the Jew ✡

Introduction

Philip N. Wheeler

Hans Nintzel added this work to the R.A.M.S. Library in 1982. It is from the British Museum printed book, 122 pages, 8905 A 15 in German "Donum Dei" (Samullis Baruch), "Abraham the Jew" (in German), 87 pages bound with 9005 A 15.

Abraham Eleazar was probably the fictitious name used by the author. It appears that the book was first published in Leipzig in 1760 with the title, "R. Abrahami Eleazaris Uraltes Chymisches Werk," although an edition from 1735 is said to exist (see Note at the end of the Forward).

The author might have been Julius Gervasius of Schwarzburg. The Forward states that the author took illustrations from the copper tablets of Tubal-Cain, a person mentioned in Genesis 4:22.

The Figures of Abraham the Jew.

From *Nicolas Flamel*, Poisson, 1893.

RABBI ABRAHAM ELEAZAR A VERY ANCIENT ALCHEMICAL WORK

Which was formerly written by the Author, partly in Latin and Arabian, partly in Chaldee and the Syriac Language and Written Afterwards by one who remains Anonymous.

TRANSLATED IN OUR GERMAN MOTHER TONGUE, AND WITH ALL THE
NECESSARY COPPER PLATES, FIGURES, VESSELS AND OVENS
BELONGING THERETO AND *etc, etc, etc,*

Written for the use and employment of the Lover of the Noble HERMETIC PHILOSOPHY

Jacob Bern. Fran. Exkhart, 1774

NOTE: British Museum printed book, 122 pages, 8905 a 15 in German *"VONUM DE1"* (Isamullis Baruch). "Abraham the Jew" (in German), 87 pages—bound with 8905 A 15.

FORWARD

ACCORDING TO ONE'S RANK, HONORED AND INCLINED READER:

What Johann Waich writes in his Commentaries about the little peasants p. 176 in the following words: "There are in these days so many arid excellent Writings of the true Philosophers strewn abroad everywhere by the Press, a great number of the same will be also secretly retained", even that can also be applied to the present Book, it lay so hidden till now in Libraries as a remarkable treasure, that only one in a hundred scarcely got to see it, and there was also paid for the transcript thereof 100 and more Thalers.

Even the Frenchman, Nicholaus Flammelus, himself an Adept according to sufficient proof, Borelli in dictionario antiquarum vocum gallicarum p. 158 has himself declared, like the celebrated D. Petraeus in his neat Preface about Basilius Valentine's Chemical Writings, not to be vexed, that in order even this gilt book of Abraham the Jew to rightly understand, to have travelled 21 years about in the world, certainly so will others no little benefit have to hope for from an understandable reading of the present from its dross, out of different complete Manuscripts with most possible diligence and pretty Work, how far they otherwise are not too much prejudiced by damaging Sophistrys.

For the praiseworthy Author of the Hermetic Triumph rest upon p. 213 Seg. that the mentioned Flammelus protested that he would have been himself deceived in his operations, if Abraham the Jew had not put him on his guard.

Yea, even Flammellus confessed in his Chemical Works, printed in 1681 at Hamburg, that he never would have found the right vessel of the Philosophers, if not the Jew Abraham the same had marked down with his proportions, the fire in which was a great part of the secret. What otherwise Ludewig Orvius won from even this book, were all those which had success, which also the seldom to be obtained Mstum de occulta Philosophia sapientum et vexatione stultorum, somewhat more exactly to glance over, knew to boast of without diffuseness.

Particularly the Author praises extremely this Jew for the curious research of some Minerals, Animals and Herbs in Chapter 3, pages 21, 22, 23; how, where, arid when the prima materia of the lapis philosophorum may be got, and if he even does not explicitly name it in that place, but, only this present Work of uncovered things, is called a rare and secret Manuscript, so will I, yet such must give a full approbation, that, this in the first part page 12 to 14 is found an expression using almost word for word, withal, also had successful experimenting with good guidance.

This Abraham had his PRINCIPIA out of the copper tables of Tubal-Cain taken almost singly and alone, as he in the second part even announces openly, p. 1, 8, 45, and 75.

The doubtful question about the Tables will not be discussed here. Who understands the original language can convince himself of this in Genesis 4:22.

This however is to be noted that Tubal-Cain transmitted his knowledge for his descendants in the mentioned Tables not with letters, but for the most part through certain Characters and pictures.

Which concealed method almost all Philosophers avail themselves of, as for instance, the Tabula senioris of the so called Rosarii allegoric Pictures.

However this Jewish Philosopher Abraham Eleazar produced all the Formas Hieroglyphicas of Tubal-Cain and explained and illustrated them through the present work very beautifully per commentarium.

Whether it allows itself also to be inferred from one and other circumstances that this Rabbi Abraham Eleazar must have had in the midst of Jewish blindness a yet passable knowledge of Jesus Christ, as the true Messiah, so is it also not to be denied that he according to his hereditary Jewish rudeness mixed in, not alone many audacious expressions, strange fables, vain consolations, strange prophecies and idolatrous Prayer forms, but also sometimes very great errors here and there. An upright Christian however will himself on nothing strike, but with Cajo Plinio Caecilio secundo Lib. 3, p. 165 press nullus est libertam malus, ut neu aligua parte prodesset. The lover of the noble Hermetic Knowledge will meet in this present rare manuscript indeed very much to their advantage, (of this otherwise diligent Jewish Doctor) when they will read it through after instructions of the Holy Church teacher Basilii magni.

Quemadmedum, says he, homil. Ad Adolesc: qui e graec: authoribus utilitas capienda, in roseto flores decerpimus, spinas relinquimus, hic profani

authores sunt tractandi, ut utilia decerpamus, a
noxiis nobis caveamus.

Herewith good-bye and remain favourable to thy
service offering.

> Your Devoted,
> Julio Gervasio, Schwarzburgico.

NOTE. The preface in the German edition of the year
1735 is more extensive and has a frontispiece of a
Jewish Priest holding a Flask.

A quote of Ali Puli in a footnote says the book was
possessed by Cardinal Richelieu.

The Book of Abraham the Jew

GOD WITHOUT BEGINNING AND END

Abraham Eleazar the Jew, a Prince, Priest, and Levite, Astrologer and Philosopher, sprung from the root of Abraham, Isaac and Jacob, wish my brethern, who through the anger of the Great God, lie scattered here and there throughout the world and are caught in servitude, much success and happiness in the Name of the soon coming Messiah amd of the great Prophet Elias, who has already prepared all his brethern. *Deni, Adonai, Bocitto, Ochyache 60 F.,* therefore wait in patience until the Hero comes. *Marantha* however, over all and sundry, not of the tribe of Judah, who receive this book into their hands, that such must be destroyed and perish, as the company of Korah, Dathan, Abiram and perish or

vanish in the Fire.

      I. N. U. C. XI. *(INC CXI)*

AZOTH: לָגֻ֞  שָׁפִים

Book II ESDRAS, Chapter 14, verses 39-47.

And I opened my mouth and saw there was reached to me a full basin, this was full as it were of water, its colour however like unto fire, and took and drank, and as I had drank out of it, then was my heart troubled in the understanding and in my breast grew Wisdom, then my spirit was got with good remembrance, and my mouth was opened, and was henceforth not shut.

The Highest gave also to the 5 men understanding and they wrote that which was said in the night, and what otherwise was above their comprehension, and they did not know.

During the night they ate bread, I however spoke over the day and did not keep silence also during the night.

There were however during the 40 days written 204 books, and it happened as the 40 days were over, the

Lord spake and said, the first books which you have written place openly that they be read by worthy and unworthy. The 70 last however keep, which thou for the wise of the people are entrusted with, for in these is the Wellspring of Understanding and a Well of Wisdom and a River of Knowledge and I did so.

DEAR BRETHERN

The Need and anxiety has mounted high, and the burden so great, that they nearly almost all crushed us into the dark grave, where all consolation ends, because we must complain day and night, that the Lord forsakes us, arid in His righteous anger banishes us from His Sanctuary, which wast', and the heathen were called in for abuse, and we be destroyed here arid there with mockery and scorn, under pure pressure lie under servants of the heathen and unbelievers, and must endure from such all troubles, so contrived. As there is nobody, who would be zelious for the law and the Sanctuary.

Therefore, dear brethren, have I thought on you a little again to raise, and to give again a consolation in your oppression; even to give a pattern, and as yet again help you, and raise up, and when the Hero comes set at defiance your enemies, and the soo coming Messiah with joy triumphant be able to receive

who will lead you out through his predecessor and messenger the Elias, then will all your enemies lie to the footstool of your feet.

Then shout thou forsaken one, that those who oppressed thee, those will be destroyed with Fire, those however who did you good, will submit themselves to you at the time.

Guard yourself, however, that you do not engage with such, for they must bear the load of the Lord. Make yourself however liable for servants and maids; separate them among you, in order that their seed cease and die out. Yes they will be your wood and water carriers according to the words of the Lord.

Therefore have patience and suffer all oppressions. Avoid however also and shun all vice. Hold to the one God of Abraham, Isaac and Jacob, and think how often that the Lord has freed you from your enemies, that, where you hold to him that finally your Heartbreak came to him, that he freed you from the disgrace of the slanderer, because they say in Psalm 95 v. 2. Where is their God? Who does not recognize you. He will not otherwise be able than on his Covenant to remember, according to his promise, and a deliverance from Salem send; the Messiah will destroy and drive away, all the kings of the heathens before us, arid conquer. Then will we with him reign, when he will then gather us from all the 4 ends of the World.

He will become our King and Messiah, be praised who comes there in the name of the Lord. Hoseannah to the King! N.B. In order that you now in such oppression may have a consolation, so take yet the words of the great Prophet Jeremiah, to you in Cap. 5: v. 1 seg. with groans in your hearts, when the same in his misery furnishes, Lord think on us and our disgrace! Behold and look how it goes with us! Our inheritance is lost, and to the strangers becomes a portion! Our houses have they robbed us of, our mothers have become widows, and we forsaken orphans, who without father, and when we drink water and would have wood, we must pay br it. We are despised, and tired from care, yet is of the driving and anxiety no end. Ah! Our sins, which we commenced with our fathers press us. Our fathers are vanished, and we must yet bear their burden. Ah! The heathen press us about the neck, we are obliged to serve Assyria and have no salt to our bread. Ah! The hero bends not, when he in his anger grows furious, so spares he not to the 3rd. and 4th. generation. But then will he cease, when he will see that servants rule over us, he will rescue us from their hands, that we must not eat our bread with tears, as it were a theft. Ah! That the sword shall not eat up our enemies, for we are few, and must die of starvation, and become as the dead.

Our virgins they make servants and deflour such before our eyes, our old men and princes are wretchedly by them murdered. Our young men are put to death under the burden. The boys have sunk under the burden of wood. Also the place is no more existing, where our ancients sat, our harps have disappeared. Our head stands uncovered. The crown has been stolen from us. Our eyes become dim, because our heart is full of care. Ah pity us, that our enemies are so great. The foxes run over the mountain of Zion, which by the heathen there must lie waste. Ah! Lord! Who eternally remainest and whose Lordship is without end, thou will even now not quite forget us. For we have become a derision to our enemies in all 4 ends of the world. Dost thou not hear our cry? Wilt thou not eternally forsake us? Lord bring us again to thee and to our brethren, in our home, in order that our days be renewed with the age. Ah! Thou art too angry with us! Remember us yet, and give us compassion. We must bear disgrace which our fathers did wrong and deserved, and made themselves guilty, wittingly and unwittingly, from one tribe to another, thy anger is even so terrible that it will not cease.

Ah! Think yet on us again, and unite us with our brethren, who through thy hand are led out and rescued from Assyria. Now Lord remember, that we are thy property. Thou wilt hear us in order that from the enemies, to whom we are delivered, the haughtiness will be taken, who calumniate Thy Name and say: Your God has forsaken you, we will badly trouble and plague you and are regarded as dogs among them. Fulfill the prediction of the Prophet Zephaniah C. 3. V. 12 seg. which he calls in thy name. I will let thee remain a poor and downtrodden people because thy Lord thee will call. Then the wicked will cease with the remnant in Israel. Your mouth will no more be false and treacherous, without fear will thee find Rest and delight, then will Israel rejoice and the daughter of Zion rejoice, that their Saviour and Rescuer has come, who releases them from their enemies. Then the chastiser will cease. The Messiah as thy King will than be with thee. Fear thee not Zion and let the hands not be tired, till you are again gathered in Jerusalem, for God thy Lord is with thee as a strong Saviour, he will again be friendly to thee and rejoice over thee, because thy misdeeds he has forgiven and thy sins sealed up. Then will the Lord collect all the rest, as a clucking hen her chickens and one will hear the sound of the sack, but in all ends of the World. Then will thy tormentors desist from their laws, and thy disgrace will be terminated, for I will end it with the task-masters and release thee from the tormentors, and will all the lame collect from the exiles in all lands. I will make you before the enemies eyes be praised and honoured among all the people on earth. With power will I lead thee, says the Lord. The Prophets Malachi and Elias will I send to you at that time, and will gather you, ere the great day of the Lord comes for judgement over all heathens. Then will the children seek the heart of the father, and I will turn the heart of the fathers again to them, before the time comes, that the earth with the excommunication is smitten. Time is soon at an end. Dear Brethren suffer it, for it

will not be long, the deliverance is near. In order
that I make the beginning, therefore a consolation
to you have I made possible, so have I here written
down the secrets of our Fathers in order that the
tribute you gave to the Romish Emperior, and you
also under the bond of servitude can help the poor
prisoners be released for the praise of the Holy
God. Therefore will I teach thee the preparation of
the Metals in Asophol and Diana, thereby with
certain words and figures, of such a pattern, in
order that with your hands you may lay hold of it,
and how you shall prepare the FIRE of the Lord,
which was so lost, when the time comes to trouble
your enemies therewith, that for your protection
have it at hand, further will I show you the place
where our fathers walled up and buried the secrets
as Jerusalem was laid to waste by the Emperior Titus
Vespasianus, in order that you may tell your
children, and also can show, for no heathen will
find the place, but only the brethren.

The marks stand to this hour, that even a blind man
among you will be able to find it, but only when the
great Prophet Elias will be present. For if you have
likewise all marks you will get it not till this
time comes. Some, it is true have already gone out
of their wits, and have sought and found the place,

but at opening of the same has fire sprung out, that they have partly perished.

Therefore guard yourself, that this Book come not into your enemies hands, that the anger of the great God be never more raised over you.

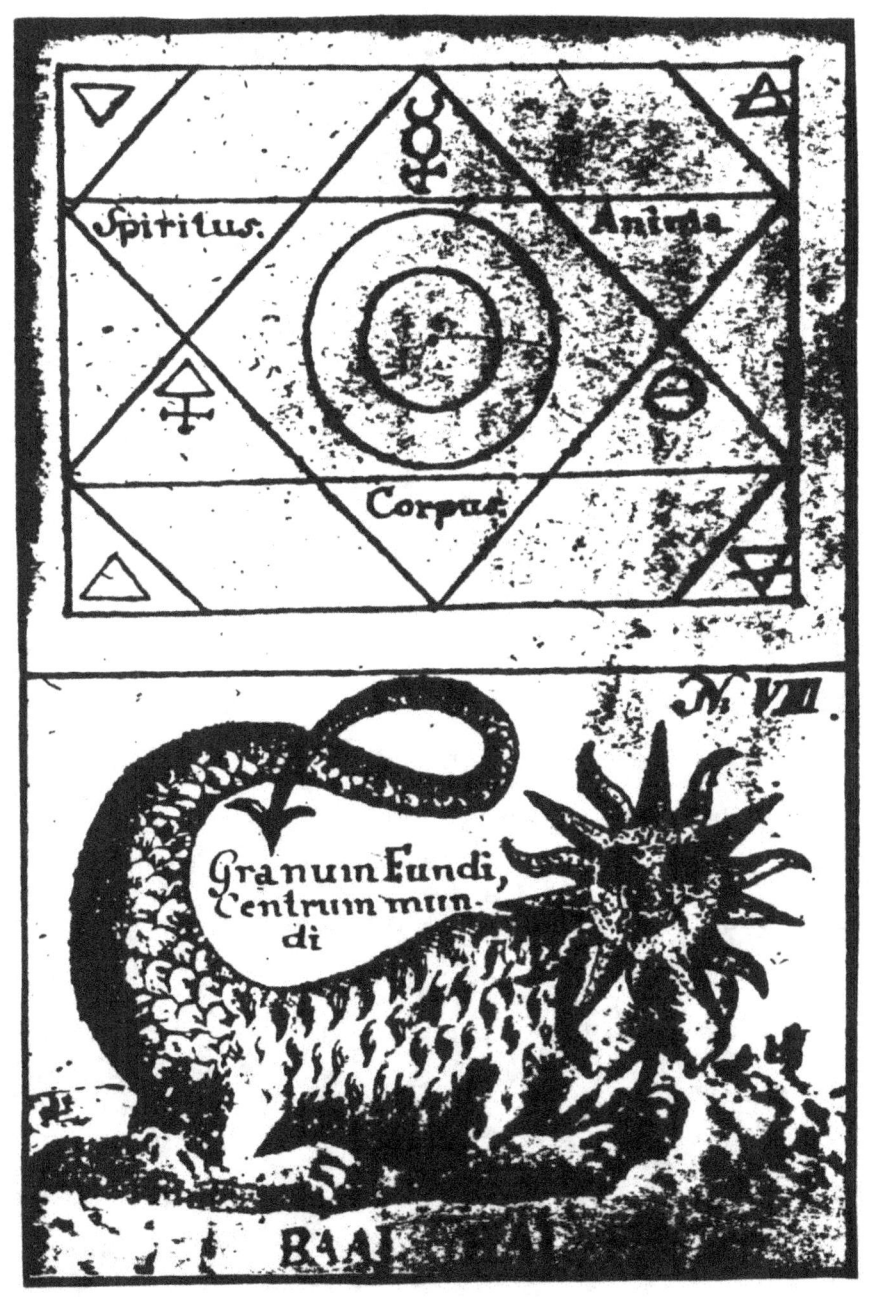

Spiritus. Anima.

Corpus

N° VII.

Granum Fundi, Centrum mundi

BAAL

Number 1

Thou lettest the Heathen say, where is your God, who will help thee? Dear brethren, therefore are we troubled to say, for we have become for a scorn among the Heathen, but the Lord hears us not, but he has bloody Vengence on our enemies, who make a Murderhole out of the Sanctuary; that we also must be expelled from our inheritance. But the great Jehovah will not eternally retain anger, but will gather us shortly to receive our inheritance again. In order that you now when the time comes to be informed where our Priests and the principal secrets of the time have concealed away, as Titus Vespanianus the Tyrant, who devastated and burned the holy City and the Sanctuary, so mark. There will such be found at the entrance into the Holy of Holies towards the East, there is a vault 500 spans deep through a narrow entrance, and is covered with broad hewn thin leaf like stones, two knee joints deep EARTH thereon, then again leaf like stones.

This is yet till this notice concealed, and will at the time when Elias with the Messiah comes, be found. Therefore dear brethren sigh and groan with desire after such; for at that time your enemies everywhere will be extirpated.

There have been some brethren who received news of this secret from their fathers. These have ventured, and have there gone in order to help their brethren. This was a good intention only; because such were not from the race of Judah and also the marks they did not understand, for they began without instruction of the same to work, they found it not. Had they understood the marks, so would they have found the concealed, if the God of Abraham and Isaac had not held his hand over it. In order that you now of the Figure and exact direction learn that which you in the seeking would find, certainly, so have I, as I received them, drawn and would discover from my fathers: Lo when you come to the Place and find the entrance into the same towards the East on the right side, so clear a way, there will you find a stone, which two knee joints deep lies marked also שׁ א lift this up, it lies 200 spans deep, should the entrance have collapsed must you further search again, till you come to that place, there you will find all secrets, which our fathers possessed and from us you shall require. Then at this time will your enemies begin to badly plague you; but be comforted, you will then get the sword in your hands, that you may fight your enemies. In order that you also have a consolation, till the time comes, that you can come and help to loosen the poor imprisoned Brethren out of their bondage, so mark, what these before written figures show to you. You should know, that God the All Highest has promised a blessing, and will give it to you, that you should enjoy the marrow of the land and drink the dew of Heaven.

For our FATHER says:

Pater ejus est sol,

Mater Luna,

Ventus portavit illum in ventre suo mari

Sal ⊕ nostrum in mari mundiversum,

Sans ⌒̲ aeris, invisibilem, congelatum.

Coelum, nostrum ▽ im manus non Mandefacieientem.

For the Spirit of the Lord is unfathomable, It hovers in the Air, it means the winged serpent and penetrates Men and all Creatures which are created on the Earth.

The winged Serpent points out to the *SPIRITUS MUNDI UNIVERSALEM,* and penetrates all things under the heavens. This is our Materia, so have we also of the coagulirten Air Repariren.

This is the **SPIRIT**; thus out of the Dew is drawn out, and with which our *SALT* is prepared. The undermost serpent however denotes our Materiam, everywhere to be found; it is earthly and also heavenly, then it is the right *EARTH,* Virginea et Adamica. That one however may know, what it is, so is such to be found under the Vegetable Roots.

This possess the SPIRIT UNIVERSALEM and is neither animal, Mineral, nor Vegetable. It is a Magnet, so it itself draws the Protmum Universi and becomes thereby a **CHAOS** of the wise Artificers.

Dear brethren mark the great secret that at all times the Perfect destroys the unseasonable, and brings it to nought, and in its nature changes, which both afterwards; veneficio caloris into an excellent Medicine can be elevated. That because their dependence is upon the Semina et forma essentialis omnium rerum a coelo stellato.

On whose account also in the metal a radius astrorum tanqua parte formal! in the earth will be generated, so have our old fathers said, the Generation of all

Metals examiniret, which form they named ex $\triangledown$ per forma interram inspissata congeal, for the radii of the stars particularly SOL arid LUNA shine upon the Worlds continually and penetrate mediante aere et aqua, within the same, and come in centro ▽ together, from thence they have a repercussionem aequalem through the whole Earth and back, but in the filtering inspissiren of the water, and make a salt essence thereof, which in itself anew a heavy flowing Substance contrahirrt, so the acanus is called and is the first metal. This mark, for it will through the constant heat be driven away, which is occasioned by the stars, says Daniel and is also boiled out, and is Asophol. With this Arcanum in a form of SALT out of the mine, or also on the earth it also makes one and the same an acid Liquorem, which the metal anew into his beginning transmits, which as the first is well to be observed. Take therefore, what lies before your feet, and of yourself with your feet treat on, for *HERMES* says: Pater ejus Sol, Mater Luna. Will now dear brethren with this fatness of the earth enjoy and drink of the dew of the heaven; and with the winged serpent which moves in the earth, get so much and with another which is without wings, know to hear, in order that you can unite the lower with the upper.

36

Now hereto must you have a clean virgin **MARCZ,** and such imbibe with dew as follows. When the time comes, that all begins to grow green and becomes a beautiful meadow, full of flowers, when the heaven is pure and clear, and the open air is full of lovely odours is swelling, pretty drops steams; very easily, when the SOL rises, make 2 or 3 round holes knee deep; the grass and turf with the flowers place on the side, the earth take out, the holes with other earth fill again full, and place the turf again thereon, in order that the mat or meadow ground get no damage, so have you the virgin Marcz. The red and yellow Marcz particularly, which out of the vineyard is very precious thereto; also when you take them out of the hole they must be taken knee deep, in order that they be free from all roots, 24 hundred weight. That they however be not stony, let such on the Mat lie spread out in order that the stars can therein work. This Marcz set 14 days and nights lie in clear and bright weather, should however Rain come, cover such with thin wood chips or straw. When the 14 days and nights are over, and the Marcz is throughly seasoned, so let the Marcz be carried away in wooden vessels, and well covered. After this take a great cask, and make in such of straw and wood a cross bar (grated); and lay such below in the cask, and thereon a quantity of Marcz; pour thereon dew, or Rain-water, taken from a

37

thunder shower is very good, leave it 24 hours so standing, then make below on the cask a hole and fit a bung therein. And through it let the WATER trickle down, till it all falls down. Pour again a fresh warm WATER thereon, after 24 hours let it again run off. In the same manner proceed also with the remaining Marcz; and then pour such clear WATER into a copper kettle, till it be quite full, and lay on such three parts and seethe down; pour the kettle again full, namely that which is boiled down leave in the kettle, seethe it down again, till on the third part, and thus continue 10 or 12 times, when now at last all is seethed down, and the third part part is yet in the kettle, so pour such in another clean kettle, set it into cold sand some days so will a SALT crystallize. This take thereout and preserve well in a clean vessel, the remaining leave again to boil down to the half, set it down again and let it crystallize till all is crystallized; then continue. On this SALT pour a clean dew, in order that it will dissolve. Then filter it and coagulate it so often, till it is pure and crystallized and prepared, so have you the right SALT Marcz, which in the Sea of the World hovers and is concealed, without which nothing can be engendered and born, and also have you the fruit bringing SALT and the prima Materialis Universal Marcz prepared and priman materiam of the old Wine.

Figure 1.

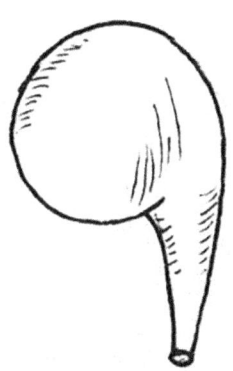

Take of this SALT 32 shekel heavy put it in a clean
vessel of **Acures**, rub it small, mix among that of
its own Marcz, wherefrom the SALT has been
lixiviated, which before heat throughly, mix it with
the Salt amongst each other and imbibe it with dew
or Rain water, place the vessel in the Sun, and when
it is dry, again imbibe, and so continue, in order
that the SPIRITUS AERIS UNIVERSAL may yet more
frequently insinuate itself therein for scarce 4
weeks, then make balls thereof, with this, fill such
a Vessel. (see figure 2). Fill this half full, so of
good Materia burnt, lay this in an oven, and another
great vessel attach thereto, in which you before put
2 measures (quarts) of distilled dew SPIRIT is
poured, lute this well, and distill through Algir
FIRE heruo termon, humor, *ALGIR,* a SPIRITUM and SALT
volatile, thereof you

Figure 2.

have the (volatile) flying serpent; thus continue
with fresh Materia, till you have from such a salt
driven all the volatile SPIRIT. Now you must take
this Volatile SPIRIT and put it in such a Vessel as
is here represented. (Figure 3.)

Figure 3.

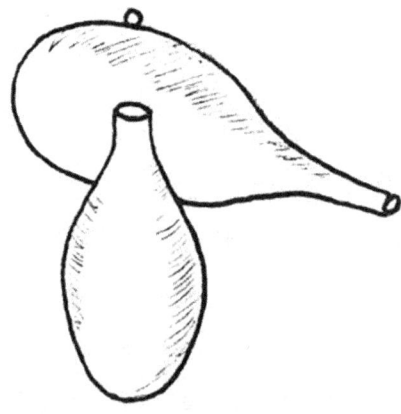

This is prepared from Acures, and place such a head thereon with a long nozzle, place the vessel in a Kettle, with WATER B.M., and distill carefully all the Phlegma over, afterwards take it out; when it is cold and place the Vessel in a *Alazabue* (Alembic) with sand filled up, and distill the SPIRIT over, this repeat, for his wings fail him yet, therefore, you must distill per Se, 7 times, till it is pure as the tears from the Eye, at the 7th. time give it its SALT Volatile in order that with it, it may rise, or lift it a part. Give finally strong FIRE so will fly the flying Serpent, and mount up in white flowers, take this and let it once again rise, till it has become pure and transparent, and preserve it well.

Then take all the matter remaining behind in the vessel, whereof the SPIRIT is driven, and bruise small, and pour distilled dew or Rain water thereon, and extract a SALT thereof; This purify so often till it becomes as beautiful and clean and transparent as a diamond; so have you the Serpent without Wings. Take care however, that you lose sight of nothing in the work, in order that the Pondus of Nature may fully remain. Take the fixed SALT, rub it small, and put it in a long glass vessel of Acures (Figure 4) and pour the SPIRITUS and your Volatile SALT thereon and shut the vessel

41

well in order that nothing thereof may fly, sit it
in

Figure 4.

a mild warmth, so will the flying dissolve the
fixed, and join; the flying serpent will eat the
fixed one, and will out of both become a fiery
creeping dragon; Here you now have the Quintuum
Essentiam and the blessing, which God the Lord laid
in the Marcz; which is of the dew of heaven and of
the fatness of the earth.

Gen. 27. V. 28, 29; and the life of all things which
are created. This Liquor is sweeter than sugar.
In order that you my dear brethren may also further
know what you should do with this blessed liquor,
mark this.

Take of OPHIRIS SOL 1 shekel heavy, make it into thin leaves, put it into such a little glass (Figure 5) and pour 4 shekels heavy of this noble liquor thereon.

Figure 5.

Place the little glass in mild warmth, so will the SOL dissolve mildly to a high yellow liquor, and a grey EARTH will precipitate to the bottom thereform. Then separate your clear solution in such a vessel that three parts of the same remain empty, therein place your Liquor, and place it, in the Name of God in the Secret Oven. (Figure 6).

Figure 6.

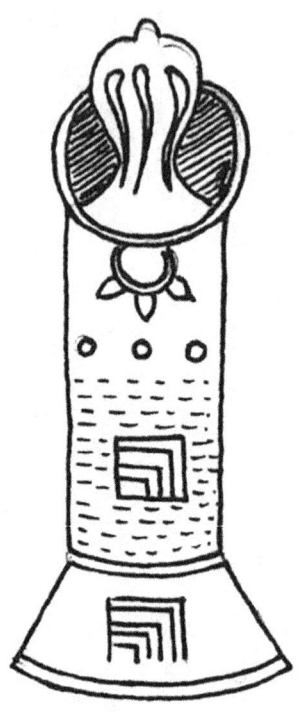

And give it a soft *FIRE ALGIR,* which should be a B. M., vapouring heat for 40 days and nights, till all has gone to Putrefaction, and become black, which the Ancients have called the Crow's head; then place it into the Ashes and give strong Algir, so that the FIRE is so strong as the Sun, when it is at its highest; so will appear the wonder of Nature, with all its colours, like to a peacocks tail, then strengthen the Algir, yet that it does not kindle, and appears after 40 days to the white, the Moon Light, and enters Diana in her snow white glittering smock. Give it yet 40 days and nights in the 4th. grade of FIRE. Here must your vessel be concealed in the ashes, when it has stood in this Grade for 40 or 50 days, so will the red Lion appear and come to pass, like a carbuncle, yet like a yolk of an egg will again separate itself. This is now the right Quintessence of which a little piece thrown on SATURN, pierces it through, and makes it SOL, for it is pure TINCTURE.

Dear brethren you shall also know, that when you dissolve SOL with this heavenly liquor, it is a strong medicine for all infirmities of the body, and as your needs require, you may make mankind wise, because the strength of the body, it strengthens, and removes mankind from all need. You shall however, dear brethren, know of all things, that this Universal liquor, before you alloy with **Asophel**, must be made metallick and effective, so to say, as the first effect to the metallic Tincture is in two ways; first with *MERCURY VIVE* purified, which through the liquor, also becomes a Water, for it has been in the beginning nothing other than a SALT WATER, and this is the right key when explained, and so place the faeces themselves, for so we proceed, as the Author teaches.

The other manipulation however proceeds thus far.

Take the Mineram SATURNI, which is nothing other, than a clear volatile seed of the SOL & LUNA; beat it small, when the same is separated from all rock, pour the liquor therein so will this blessed Minera dissolve in it. The clear solution pour off, and pour more liquor on, when all is dissolved, pour the clear solutions together in a Acures, place it in a cool place and so will wonderful crystals shoot forth; so can this wonderful salt in a different manner be brought to a Tincture.

There is yet a way to obtain this secret, namely thus: Take the purified SALT before the SPIRITUS is driven off, and make it night dry, and put it in such a vessel, whereof one distills, place the same with the SALT in a B. M. or in Horse-dung, the B. M. however must be constant, and for so long till only the SALT itself changes into a OIL ichten[1] Liquor and separates from all uncleanness.

Pour the clear from all Faecibus, into a clean glass, and place it in B. M., and distill the phlegma carefully over. There will however be very little, and keep in such a heat till it again coagulates, then place the glass again in fresh horse-dung to putrefy, and to dissolve, so will it

---

[1] *an unctuous, oily liquor. HWN*

again dissolve, and then again coagulate it again, and this repeat, till that only the SALT is fixed and in the FIRE, flows like Wax without diminution. When you now have it so far, keep it well preserved.

Take OPHIRIS SOL and dissolve it in a *WATER GAZA,* and when it is all dissolved, then distill all of it over in a glass vessel, and pour WATER thereon, and dissolve it anew, and distill the WATER again therefrom, repeat this oftimes, give at the last a strong FIRE, so the WATER GAZA goes mostly over, dissolve it yet again, and put in the solution of the Rhystone, so will the SOL draw it to itself. Distill the WATER again therefrom till it turns to a powder; place this in a closed glass in the FIRE, and so the *PHYTON* will fly and leave the SOL, as a not solid EARTH, quite open and porous. This EARTH sweeten several times with clean water, and dry it, of such a powder take 1/2 shekel heavy, and from the above liquor or *LAPIDE AIBACHEST* (Alkahest) 4 shekels heavy, rub it in a glass of Acures together, and put it in an **Alingel,** seal it up, that nothing can come in; place the vessel in the furnace, and give it Algir FIRE, till it flows together to a red stone or powder.

Dear brethren with this can you also do wonder, for it transmutes all known metals into SOL.

47

When you also have the above UNIVERSAL MERCURY, which you have prepared out of the Volatile snake and the fixed, poured on the *Albaon,* so in the end such is green.

For there lies in such the life of all metals and minerals as the right key, and this concentrated SPIRITUS MUNDI, can in all things be used in exaltations virtutia Elixiris de prolongandum vitam.

For Heaven and Earth are preserved through this. This is the right green **Alazagi,** with which one can do wonders, as I already before taught, and will show afterwards in the proper places.

Dear brethren, in order that before you accomplish this secret work and also have food and nourishment; so will I teach you some small primitive work which in all places you can do. Take melted *Almusater Alatren,* Have it and **Celuvialatel,** rub it amongst each other, put it in a strong stone vessel, and sublimate it up, and on that sublimate mix again among the remainder, and put also some of the fresh Alatron therein, and this sublimate up again, and repeat for the third time, and so you get a GOLD SUBLIMATE, which you shall also require, let 32 shekels heavy of the pure Copper Metal, in a strong

EARTHEN Vessel prepare, let this flow in a strong FIRE, when flowing thin, add the fat sublimate 4 shekels heavy to it, let it flow for one hour, then pour it out, so have you the great Secret thereby to accomplish a living.

Yet must you remember this necessary manipulation, when you have a right fat Marcz, and with dew or WATER GRANDI extrahiret so now you boil such down, so expert that in it a SALT can crystallize, and on account of the great fatness and fat viscosity of this SALT. When you see this, so let it safely evaporate to a thick liquor. This liquor is far better than the crystals. With such you can go to work, that you mix therewith so much of its own throughly heated and lixiviated EARTH, and then distill over, as has been taught, and so you get at once so much of this Volatile SPIRIT.

The fixed SALT remains behind, which can be lixiviated and goes in the work as said before. Guard yourself however that these Secrets come not into your enemies hands.

Conceal it from them as you only can, till the time of deliverance comes. Then rejoice the forsaken daughters, that the Lord and Holy Adonai has made an end to your ignoming.

Plate I.

Number 2

I, Abraham Eleazar, continue to teach you Dear
Brethren, as our Fathers in the Wilderness sinned
against the Lord through Idolatry, Moses made for
them a brazen Serpent and stuck it to a Cross, that
such would be seen by all the people, and they again
from their merited plague might recover.

Therefore know, when you can fasten with a Golden
Nail, the Serpent Phyton on his Cross, so will you
want nothing in Wisdom.

Therefore is Nature most hidden and this is the
whole secret in the Art, that we draw out from this
secret Materia the SPIRIT PHYZONIS and PIRTRE SOL,
as the Sulphureous WATER, as the strength from FIRE
& Air of imperceptible form, separate with small
trouble; therefore is it full of Spirit, and holds
in itself a fat fruitfulness.

This will now be driven out and separated, that as a
clear WATER and appears to the eye as a tear, i. e.
SPIRITUS MERCURII. This loosens the common Phyton
and makes it also a WATER, yes to an Aldibid.

That however you know and become acquainted with the Materiam; so is such our old ALBAON Abacschozdii, is a Minera, so there in the mountains is found, and such is of three different sorts. The first is in all its parts, is used by surgeons, it is right to say a volatile ore while it stands as pure Tincture or Seed, so however volatile, while all of the FIRE is taken away, till very little. When such is melted it gives a little sparks. This is best. The other appears yellow and also with little black sparks, is often found in yellow gravel or sand, and has much auriferous volatile seed. The third is grey and white and a very poisonous kind, a right SATURN, which has the power, with its poisonous breath to kill. Therefore one must be very careful when working with this, when in a dry form, to get its sweat.

In the wet way it is more sure, for when this old one is put into a bath, it betakes itself into it. Thou shalt also know that this old one is of a SALINE property, therefore such is dissolved only into a bad liquor, so from Kaly and SALT is the sapient crude aa prepared, this is through a penetrating poisonous SALT, which is a pure Phyton, heightens the bath, that this old one himself should dissolve in such, till to a little, which the Fera of the heat of the Body is, the clear solution is poured off, such is put in a cold place, and so is produced the Old One in quite another figure, in beautiful crystals, and also it is with the other 2 preceeding particularly, but the first is not so poisonous.

When however this Old One is placed in his bath, so will his body be divided and then appears its inward SALT Balsam, which, is pure Tincture, and this Materia by many, will be thrown away on the hills, and also such a materia is found in the pits, because they often throw such away, because it gives from itself a strong smell, and also often kills the men, for the 3 sorts, is already of the SPIRITUS MUNDI UNIVERSI made volatile, therefore such blows without intermission.

The first and other Materia is not so volatile, yet has in itself the SPIRITUS MUNDI buried, in this mineral essence, and has a right to say, is made a Magnet, and is inspissirt in full and free operation and in full Course to become a Metal, but not yet to a Metal or Mineral, nevertheless form a Minerali imbutus.

This now is the Materia, which the wise choose:

Materia non putative, sed vera e
experimentis comprebata, MATERIA
saltemunica, et res, ex qua, hic lapis
unice et solus absque pregrmno additamento
praeparari necessum habet.

Dear brethren, take the same, this ma a which women
colour the hair, so can you in eligenda ma a not
miss, for without the Dragon Phyton can nothing be
done in this Art. Therefore direct all your thoughts
to the Phyton. Nam est un Phythone quicquid
queerunt, Sapientes. For nothing in the world has
such a power to destroy Metals.

As the Phyton alone, But, dear brethren it is not
the common Phyton, but our SPIRIT PHYTHONIS,
although with the common Phyton, our SPIRIT
PHYTHONIS is infinitum multiplicirrt, for our SPIRIT
PHYTHONIS transmutes the common into its nature, as
it also changes the nature of SOL and all Metals,
for it is the primum Ens Metallorum, that is, the
Spring of the Ancients, the flower with golden
leaves and so from pulling and tearing the poisonous
dragon is covered and preserved.

Make our Old One a Heavenly Green SALT. N. B. And
from such a Phythonis, a living water which burns

not; place Ophiris SOL in the solvent and putrefy in order that it becomes black, white and red. With this you can vanquish the World.

Therefore dear brethren take heed to my teaching, for I will reveal here a greater secret to you, and with two ways show you how to obtain the great Quintessence.

Take our Materium Magnesiam, Plumbum Nigrum, also called Bismuth or Puch, such as comes from the Hills; 10 or 12 tit; make such an Old One into a fine powder, after you have cleansed all rock from it, and put this powder into several broad Alazabus, place these so that the LUNA can shine thereon, and the dew may fall therein, but no sun dare shine thereon, and no rain come thereto, and let it stand thus for 4 weeks, you must however move the powder round every day. When the time is past, take this powder, as our Old One, and put it in a crooked necked Acures. (Figure 7).

Figure 7.

Lay such in a Alazabus filled with sand, and that the sand covers the Acures and place it in the Oven and give it Algir FIRE, herno, termon humor, algir, lutirt, and from this great ball of Acures, that you can see how it goes, and distill a sweet, yet penetrating Spirit over, which is white, which is the doves of Hermetis. Continue heating till no more comes over, then let it cool. Take the receiver off, in which Our Old has concealed his Eva, take the distilled Spirit and mix with the residue or Caput Mortuum, and place into such a Vessel viz. (Figure 8).

Figure 8.

Place this vessel in the sand, up to the level of the liquid and distill and cohobate several times, distill this gently over in order that the SPIRIT PHYTHONIS comes over clear like a tear, and is cleansed from all impurities. Reserve this Spirit and take the remainder or Caput Mortuum out of the crooked necked vessel, and put it in a vessel of burnt earth, put it into the FIRE and calcine it strongly, then extract the SALT out of it with distilled water, and purify it; so have you the Adam, made out of the earth and prepared.

Now you must make this living, and give him his Eva. Rub this SALT small, and put it in an Alingel, and pour the SPIRIT on the dead body and firmly stop the glass and close it well, place in a gentle heat, so will Adam take Eva to himself, and become one, and so you have here in short a Liquorem Universalem and with this you can rightly dissolve the Ophiris SALT, and then coagulatione et fixatione in verum semen suri lapidem philosophorum redigirre. Dear Brethren I will not withhold from you through the help of the all highest how to prepare this secret Phythonis Liquorem; so take such a Liquorem and pour it on fresh Albaon made into a fine powder in a great vessel of Acures, shut it well, and place it in a gentle heat, and you will see a Smaragdine colour. Pour it off, and add fresh Liquor thereon, till all is extracted; pour the Extraction together into a clean vessel. On the remaining Corpus pour a distilled WATER, and extract the remaining tincture. Pour this off into another clean vessel, and distill it off (the water) and add this liquor to the other extractions, and place this in a cool place, so will a Smaragdine SALT crystallize, of great strength. Take this out (the SALT) the remainder leave in a mild heat, to evaporate a little, and again let it Crystallize, and you will get more SALT, and you will have the real GREEN *LION* of the Ancients.

This is a living SALT. Take this SALT that you have extracted from this EARTH, mix this SALT with the EARTH or (caput mortuum) rub them small, and make little balls thereof, moisten these balls with the liquor, so that nothing foreign is added, put such in a crooked necked Acures; (Figure *9)*.

Figure 9.

 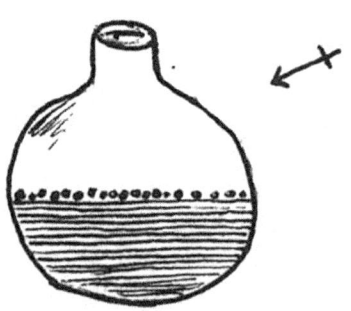

lay this in an Oven in Sand, and your vessel should be covered with sand, lay also a large ball thereon(?) and well closed, that the *Aroki* does not go through and distill oven; by Algir a flowing

SPIRIT comes over, pour this back into the vessel, (cohobate) and distill over again with strong Algir, then take the Receiver off, and preserve this well. The remaining Corpus bruise, and rub it small, and extract the SALT thereof, and purify it, and pour your Liquor upon it, place this liquor again into a crooked nedked Acures and cohobate the liquor so often, till all the SALT comes over with the Liquor, and put this Liquor into a high Vessel (Figure 10).

Figure 10.

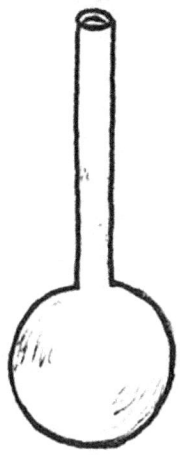

and distill all moisture therefrom, so you will find in Fundo a beautiful pure OIL as heavy as Plumbum, also pour this SPIRIT, which from its moisture you must have previously separated, upon your Oil, and draw off several times mildly, till your OIL is as thick as a flowing brook, remains behind; thus you get the Gumi of your sister Manna, and the true Chaos of our Fathers.

Then put it into a long necked Acures, and carefully melt the neck closed, put it in a mild heat, so will the Materia resolve into a crystalline SPIRIT and settles as a brown red earth. This clear WATER is the *Coloumba Dianae*. Such pour carefully from the EARTH and rectify, and well preserve it, i.e. lutirt that you clean it from all impurities, and free it from all Phlegmata. Let this flow over 7 times, so have you the prepared Phythonis. This now is the life of all things. Preserve it as a great treasure, for this Bird does not willingly let itself be seen by the Godless of the World.

Take the humidity or Phlegma, and extract out of our EARTH the red Pirtre or fiery red Lion. This solution pour all together, till nothing more is extracted, and distill the Phlegma therefrom, so remains in the vessel a red glancing Materia, as a Blood and the right flowing Ophiris: SOL of the Ancients, the Blood of the Dragon, take this and preserve it well.

The remaining Faeces take all together and calcine and extract the SALT of splendour. Clarify it that it becomes like a diamond. This SALT rub in a clean vessel of Acures among the Blood of the Dragon, and pour the Columbian Dianae thereon, close them together and put it in a mild heat, let it stand, till they are all united, so you get the right lac virginis of the Ancients, with which one can do wonders.

When now all is again in a Liquorem, mediante circulatione redigirt, and take of this Liquor 10 shekels heavy and of the Ophiris SOL or LUNA 1 shekel heavy made into thin leaves, and put it together in a Alingel, close it hermetically, put it in our Oven to stand in digestion and circulation, till all is dissolved, then let it stand in Algir FIRE, till it has gone through all the colours, so you will get a blood-red glittering Carbuncle, a very great Medicine, which is unsearchable.

This is now a Ferment. Melt 4 shekels of Ophiris SOL, and let it flow in a good vessel of burnt EARTH, place upon this 4 shekels heavy of the Tincture, and it will penetrate the SOL, and change it into a blood—red essence. So is the King of the Ancients and the right Red nature born.

Of this will 1 grain, change 1 shekel heavy of other metals into Gold. Or when you add LUNA, in the

conjunction in LUNA change. Then mark, when you add
the LUNA so you get a stone or Tincture on LUNA,
also a great Arcanum in Medicine, and the Tincture
will appear violet glittering.

Dear brethren, the Ancients have not had one Way in
preparing of this Mystery. For some, when the
Columbam Dianae they had separated from the Chaos,
they have allowed it to fly 7 times, in order to
separate from it all Phlegmate.

Then you have taken this, and a purified Phyton,
poured into a crooked necked glass, so as the Phyton
is eaten by the other, then drive it over again, so
that only the EARTH remaining, which throw away.

There have some that also raised, and have increased
their Columba Dianae to infinitum, for if you
distill it several times it becomes stronger. Then
have you out of the red EARTH, which WATER Adamicam
called, the soul, the Life, the FIRE of the
Ancients; yes the red fiery Lion extrahirt, the
Phlegma distilled off, till it becomes a glittering
Blood which they name the Blood of the Purple Snail.

This is to be well preserved, then take Ophiris SOL
in leaflets and place into an Alingel, pour on the
Columba Dianae so much, until the gold is dissolved;

to this solution pour the Blood of the Dragon, and
have brought FIRE to FIRE, for this is most hidden,
and without this in the work, you could not in an
Eternity boil out, so has this Liquor become as
Blood. Then take the faeces, and strongly Calcine,
and with the Phlegmate extract the SALT, and purify
it, till it becomes clear, and like a Diamond, then
rub it small and put it in an Alingel, and pour on
so much of the gory Phythonis that the SALT will
dissolve. The glass must be Hermetically sealed, and
then placed in the FIRE, and allowed to go through
all the colours till the Quintessence has become
firm, and then take it out of the fire, rub it small
again, and sprinkle it with the fiery gory Phyton,
and again Hermetically seal the vessel, and allow it
to go through the Colours again, and then have you
found the augmentation, which can be practicirt in
infinitum.

Then take of this Mystery, when it has become fluid,
and have a Ophiris SOL purified and melted with Puck
Bismuth and in the fire allowed to fume.

This is the King, which is eaten by the Wolf, and
again Vomited forth. Place 8 shekels weight of this
purified Gold, and when it is flowing, throw in 2
shekels weight of the Tincture, so has the SOL
become a clear Tincture. Of this Tincture add again

8 half ounces (of SOL) and mixed together, place this mixture into a vessel, and pour on the fiery Dragon, place it in the Oven and allow it to go through all the Colours, and continue this till the Tincture again becomes fluid, then take this out of the Fire, and add a part of this Tincture with melted SOL, until all the fiery Liquor is consumed. Then take this Tincture out of the FIRE, and add an equil part of melted Ophiris SOL and then of 1 more part is necessarily employed, and then another part is added thereto. And then the Columba Dianae is thereon active.

N. B. Others of the dear Ancients have before in the Columba Dianae dissolved Ophiris SOL, and have fixed and activited this Liquor therein, which is better, and the Work is increased in infinitum. Of the Tincture add 1 grain to 16 shekels weight of the Phyton thrown into the FIRE, and so it becomes the most lovely Ophiris SOL.

The more times the Tincture is reiterated the higher it becomes, even as you have assisted in the propagation, and so it has become a great Treasure.

Dear brethren; the dear Ancients have gone another way, that after you get the Green Lion, as the SALT of Nature, so have they taken such a heavy SALT, and in the high pot (Figure 11) they have placed it, and closed it well, and given it a moist steamy heat, or placed in Horse-dung, and let this secret SALT stand

Figure 11.

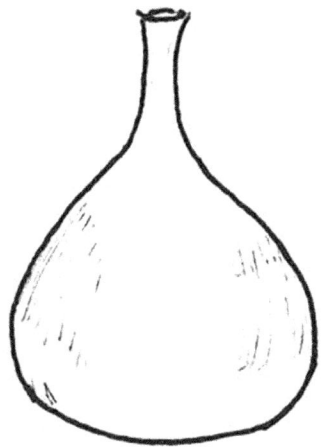

so long until it dissolves itself into a Green Liquor, then they have poured this into another high glass, put a head thereon, with a nozzel, and hung it in the Ashes to distill, till the Columba Dianae began to flow, when it was all distilled over, they poured it back, (i.e. the SPIRIT) thus distilling and cohobating till nothing remained behind but a brown EARTH.

This afterwards they extracted the Phlegm from the extracted SPIRIT, and so obtained a red fiery water, this have they put with each other in a clean vessel and distilled the phlegma thereof, till there remained a blood red DUST. Out of the remainder a SALT is extracted, which remainder is first calcined, the SALT clarified, till it became Crystallized, so have they the new EARTH prepared, then they proceeded in the rest of the work as before taught, and got the great Mysterium.

Some of the dear Ancients have our Old One a bath prepared from the SPIRITUS MERCURII as formerly taught; and such with its own moisture (blood) dissolved and closed.

Then have they prepared a SALT from such, and from such a volatile SPIRIT a blood red OIL is driven, further they have taken the pure EARTH, in which the Brilliant SALT was concealed, and they poured the red Liquor thereon, and distilled and cohobated till all passed over.

Then have they put this liquor in a clean vessel, and dissolved Ophiris SOL in it, the EARTH they have taken out of the above vessel, and this Liquor poured on, only one time, this they have distilled off per Algir, and again poured thereon, and this so often continued, till nothing more rose up, then have they poured a fresh Liquor thereon, and then continued the work therein till all the Liquor remained together, fixed and flowing in the FIRE like a Wax.

This blood red salt have they alloyed with Ophiris SOL, and melted with each other, so has it become a Tincture, with which they however tinged the other Metals, and have transmuted them into SOL.

Dear brethren our Ancients have yet further seen and taken the Universal Key, whose preparation likewise in the beginning was taught, prepared from the SPIRITUS MUNDI UNIVERSAL.

This liquor have they poured on our Old Ones, which they first beat small, and placed each other in a Vessel of Acures, in a small heat, till the Liquor is tinged to a grass green colour. The Solution they have poured off, and another poured thereon, till all is extracted, they have poured the clear Solutions together, and distilled 1 part of the Phlegmitis SPIRIT off, and placed the vessel in a cool place, till a pretty Smaragdine SALT was crystallized. The remainder they have taken and distilled one half of it off, and placed the remainder in a cool place, then is there more green SALT crystallized, and the remainder they have reheated till all the SALT was crystallized. Then have they taken this SALT as a great treasure, and this is in truth the right Green Lion; for out of this NITRE is a SPIRITUS MERCURII prepared with a blood red OIL of great effect, and the dear Ancients have prepared in this way the great Mysterium, then what concerns the after work, so is such all one work. You may also in the wet way choose the one you wish. N. B. The after work in the wet way is the Menstruum Universale.

Take putrefied URINE and the acetum distillatum SPIRIT OF TARTAR

$\overline{\overline{aa}}$ and putrifie with each other, then put CALX VIVE in a high alembic, put a still thereon, which has a hole above, and well luted.

When this has happened, then put the Alembic in sand, and give FIRE, now when you think that the Alembic is heated, pour through a filter of the mixed Liquor, through the hole in the still, so will the CALX VIVA be heated, you must however close the hole firmly and must not pour too much on the CALX VIVA at once, otherwise it would run over, (pour it on gradually) then will thy SPIRIT go over, collect it, and rectify it yet once per se. With this you can extract from each Minera its Anima, whenever you wish; the menstruum can be prepared without FIRE, but the CALX VIVA must be good, and is also better in the Operation, when one will prepare this, so instead of URINE take a strong SPIRIT of ᚜ᚘ ; so also must everything be well rectified and clean, then it is practicable.

With this Minstruo extract the life and SPIRIT out of Old Albaon, arid according to instruction work in the Wet Way.

Also have I again, dear brethren, shown a way, that you can assist, and come to the help of your poor brethren, who are otherwise in need and misery. Forsake not the poor Widows and Orphans. Be very reticent. Laud and Praise the Name of God, His Holy Name JEHOVA. Cry, that the air resound, HOSIANNA, to the Son of David our brother, Our King, Our Deliverer, Our Saviour.

Divide Them.

Plate II.

Number 3

Dear brethren, one might well wonder, how our dear Ancients hit on the thought to prepare a MEDICINE, which has alone happened through the inspiration of the great JEHOVAH, that the Metals could be ripened and transmuted higher, and how therewith they went to work, that they found the PRIMUM MATERIAM.

Dear brethren, one says: TITULUS INVENTIONIS may be often POTIOR PARS INVENTIONIS.

As they saw that SOL and LUNA were generated by so long boiling in the EARTH of Nature, so must we freely wonder, how yet the dear Ancients fell on the thought, that through Art and the help of Nature, they could renew the way, and bring to pass by the EARTH, SOL and LUNA in such a short time.

Alone the dear Ancients have learned through diligent speculation of Nature, what was their origin before the commencement, and out of WHAT the LIFE of all creatures took its origin, and how the Generation and commencement of the Metals may have their Origin. This the dear Ancients have very closely observed, examined and found, that all metals took their origin ex MERCURY or PHYTHONE;

thence are they driven, and have taken the PHYTON, and have also therewith gone to work.

They have however experienced harm with such because it has become too metallic; so have they sought another MATERIAM, in which the SPIRIT MURCURY is not yet made metallic; and found such likewise among the Minerals, and afterwards called this with divers names, as Albaon in ARABIC, in Latin, Plumbum Nigrum, Abackhozodi, a black heavy stuff; they have also called it MAGNESIUM, BISMUTH.

Our Ancients have called the stuff PUCH, it is however not the common STIBIUM, but a black grey rock, often with white and other beautiful colours adorned, heavy in weight, as I already taught you.

Will you now dear brethren go by the old road, so will I teach you shortly and plainly, not with many Names, or in figures, as the Egyptians left it behind to their children, which are not to be divined or unriddled, but clearly and brotherly.

Formerly I have pointed out and shown to you the wet way, and how you can find and prepare in such a way the Mysterium, and such a way is without danger.

This dry way, that I will describe to you and teach
you, is somewhat dangerous, yet if you follow my
teaching, then will it not fall heavy on you, for as
I have described to you in all my Figures
throughout, two ways, so to the intelligent it is
not difficult to understand, and have also wished to
show in this Figure; for here you see flowing from a
desert a LUNAR white WATER, which is from the old
progenitor of all things, prepared in two ways:

Firstly however, you must understand, what of the
two ways is taken; namely the first proceeds from
the Fatness of the Earth, out of the PRIMORDIAL
CHAOS. The other from our black heavy and white
lump; that however the serpents crawl in the grass,
and is of divers colours, the PHYTON in the dry way,
for this PROMTUS is very poisonous, yet some times
it ascends in the hills, and so becomes a Flower,
nearly medicinal, whilst then it is not so
poisonous.

Dear brethren, take our hard lumps, and make into an
inconceiveable powder. Mix this powder with some

small broken stones $\overline{aa}$ so that in the
sublimation, the powder does not melt, place this in
a Pot, as you see, and have a small Oven. (Figure
12).

73

The pot must be of good burnt earth, on this put a great glass head, and place a vessel before and give FIRE, yet that the Vessel is high up in the oven; when you see, that no more steam rises, then open a REGISTER, Mark, you do not lute the vessel, for otherwise it would be dangerous. When now all the vapour is gone, open the other Register, then will the white flower mount up, or

Figure 12.

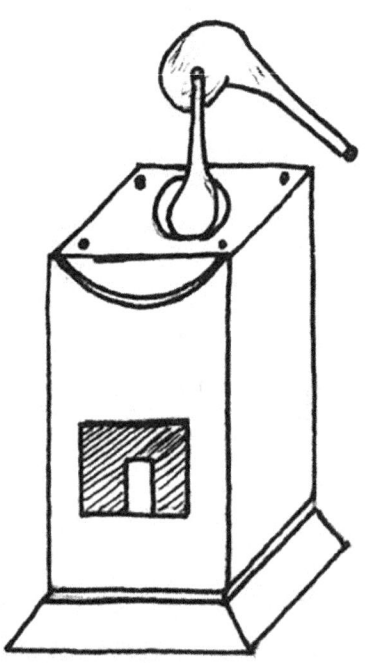

the Bird of Hermes, as the Egyptians call it. Stop the FIRE, yet take heed, that the materia melts not, thus you should before mix something among the MATERIAM, that it does not melt.

Should the head become full of flowers, you must remove this head, and replace it with a new one, and thus proceed till nothing rises up, take the flowers from the heads and preserve well. If there goes a little Liquor over, wash out the flowers (from the head) with this Liquor, for you have its life in such.

In such a manner you can make as many flowers as you wish, and when you have a pretty tolerable portion, gather them together and preserve them.

N.B. I say again, guard yourself, that you do not destroy the flowers with too strong of a heat.

Take again a fresh MATERIAM rub therein all your ascended MATERIAM, place this in the vessel, and place the head again thereon, give the FIRE carefully, then will more flowers ascend, when the head is full, take it off. These flowers can be pre-pared also in this annexed vessel, that one puts several heads, one over the other, for otherwise, when the head is full, you must take it off, and put another thereon, which causes much annoyance in the work, and many flowers are lost.

Continue this till all the flowers have ascended, then bring all the flowers together into the distilled Liquor, put fresh MATERIAM into the vessel, and sublime the flowers, and so continue until you have released the PHYTHONE from 12 tit of the Materiam: i.e. when you have the flowers from 12 tit of the MATERIAM, you have enough. Then from our ALBAON ascends the most, the remaining MATERIAM preserve well. Take all the flowers together, and weigh them, when you have 1 tit in weight, proceed thus.

Take LUNA and shut it up in a strong WATER prepared from Kalti potash and SALT EARTH. When it is thus shut up, then pour Water thereto, in which common SALT is dissolved, then will the LUNA be precipitated, which sweeten well with WATER and make dry; take of this 16 shekels heavy, rub the precipitated LUNA with 1/2 of the flowers (i.e. 1/2 tit) and being well mixed place this in a vessel of ACURES, place in the Oven, as above mentioned in Sand, and heat it carefully. (Figure 13).

Then will the flowers ascend much prettier and clearer than in the commencement, and a SPIRIT lisches WATER will go over. Take your sublimed flowers, rub them again among the remainder, and sublime it 4 times more, then it becomes like a diamond, and is not so poisonous. The remainder can be melted and purified and thus recover the LUNA that is left; or you can sublime the Columba Dianae 2 times more, then will they get a great brilliance; this take up for it is now already MEDICINAL, and no more poisonous. The other half of the flowers, as you above preserved, take and make with the WATER GAZA out of the Orphiris SOL, a DUST, which sweeten well, rub it among the flowers and bring it into a vessel, and do, as you were shortly before were taught, and let this rise also 4 times from the FECIBUS, then let it rise two times more, then have you the PHYTHONS golden wings prepared, and then is ready the noble flower CHELIDONIS; then have the Ancients taken the RESIDUUM, and burnt to ashes; out of these ashes they have prepared a SALT, which was clear and clean; this they have rubbed small, and weighed; with six shekels heavy of this SALT, they have rubbed 2 parts of these flowers, and put in an ALINGEL, and placed this in an Oven in Sand, and let it stand in ALGIR FIRE, and in a short time has the Balck One appeared, and then it goes through all the colours to a White Stone. This have they taken out, and rubbed again 1 part therein of the COLUMBA DIANAE, and again put it in the Oven.

Figure 13.

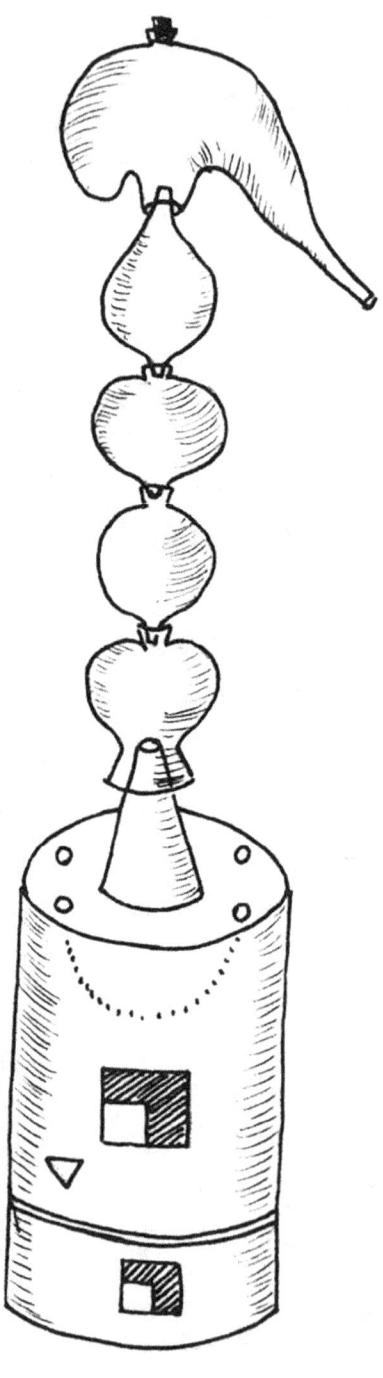

This they have repeated 7 times, then they have let flow four shekels of LUNA, and then thrown into it 2 shekels of the Tincture, and let it flow well, and so have they got a Tincture to transmute other metals into LUNA, and 1 grain will transmute 12 to 16 shekels of other metals into LUNA; Also they have proceeded with the golden flowers, and have gotten the Red Tincture, to tinge other metals into SOL.

Dear brethren, the Ancients have seen and found yet another way; they have driven up the flowers from its own CORPUS 6 to 7 times, till these flowers have become glittering and pure. Then they have taken and weighed them, and rubbed them with the flowers of PHYTHOANA ANIMA, and put this together into a vessel, and sublimed in a mild FIRE, so has one Phyton the other swallowed up, mounted out and up.

This have they taken out and let ascend alone aloft per Se, so have they obtained the Phyton of the Wise Masters, and then multiplied it. This they have named their White Flower. Which however the Ancients understood nothing, and it is not true, that one could increase with the Phythone alone, for they have taken the RESIDUUM, and calcined this to the half, and from this prepared a SALT, this SALT they mixed among the other uncalcined MATERIAM, and placed this in a good vessel of burnt EARTH, then have they melted it with each other, arid after the first part evaporated, got a brown red ACURES, this have they rubbed small, and weighed, and rubbed again 1 part of this ACURES 3 parts of the white flower, and put with one another in a vessel, and let ascend, and this have they repeated till the white flower ascended red.

Figure 14.

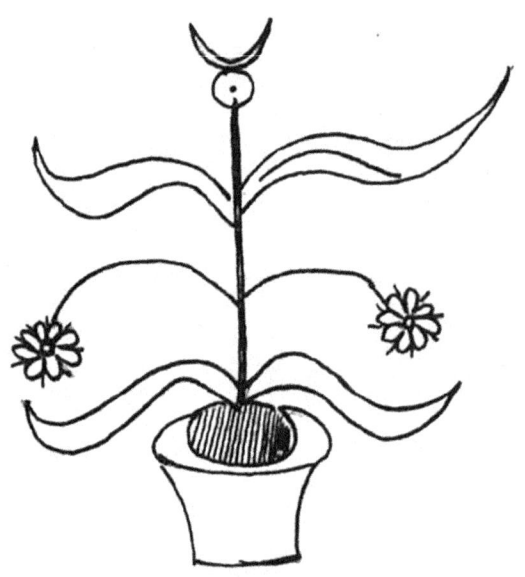

This have they named the red LION and also their SOL, for the PIRTRE out of the ACURES has jointly ascended, and this they have alone preserved, as also 1 part of the ACURES.

The remainder have they taken, and calcined with
strong FIRE, and made a clarified SALT out of it;
then they rubbed the ACURES therein, and also the
red Flowers, and put such in a stone vessel of burnt
EARTH, and put this in the FIRE, and let the FIRE
commence slowly, so has the red Flower flown
jointly, then have they continually struck the
vessel, and taken this out clean, and rubbed it
small on a POVIR, and preserved it.

Dear brethren, when you have come so far, then you
have the Tincture in the Oven FIRE cleaned and
multiplied; (Figure 15)

Figure 15.

you have taken a vessel, as I have shown, of good
burnt earth, and such put into the fire; take 4
parts of the red flower, and 1 part of the white
PHYTHONIS, rubbed amongst each other in a vessel of
AGATE or ACURES, put into the crucible, so has the
red flower spread over the white one, and with great

lustre with each other allied and dissolved, and made also your Tincture in little time; then have you taken the vessel out of the FIRE, and such laid down, and carefully picked it all out; and then again rubbed small, and rubbed amongst *4* parts to 1 part of the COLUMBA DIANAE and then melted, and thus you have repeated 7 times, and then you must cease, because your Tincture is liquid, and it penetrates all vessels; of this take 4 parts, and carried such to 8 parts of Ophiris SOL, and so it becomes a red penetrating Stone.

This you have divided into 2 parts, and rubbed 1 part among 2 parts of the penetrating Tincture and then also put therein of the COLUMBA DIANAE, and then melted with each other, and repeated this 7 times, then have you augmented your Tincture in INFINITUM; the half you reserved of this Tincture melted with SOL, you have employed for your maintenance, for 4 grains of this Tincture transmutes 16 shekels heavy of other metals in flux into most beautiful SOL.

You have also prepared your Tincture, with the white Flower per Se, or with the addition of the fabricated LUNA, the most you have only prepared per Se, and increased with the Phythone, and then have you taken Ophiris SOL, and made it into an

incomprehensible DUST, well sweetened, of this you
have taken 1 shekel heavy, and 2 shekels of the
COLUMBA DIANAE rubbed thereon, and placed this in
the ALINGEL, or into another vessel as here stands
Marked. (Figure 16).

Figure 16.

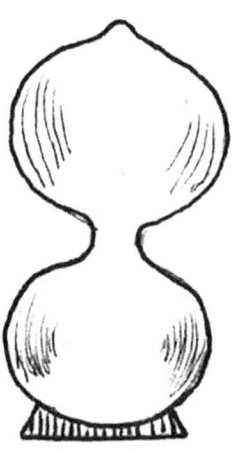

This vessel is closed well together, and place this
vessel in an Oven placed in sand, and let it stand
till the COLUMBA DIANAE has ascended, then take out
the vessel, and rub the Ascended again amongst the
Lowest, and thus repeated so often, till it has
flown together firm.

Then take this Red Tincture out, and again rub
therein 1 part of the COLUMBA DIANAE, and this again

put into the vessel and make it firm, and this should be repeated 3 times.

Then must you cease, because the Tincture has become so penetrating; take this out and weigh it, then let 1 part of the Tincture flow with 4 parts of SOL, and so the SOL will become pure Tincture; from this Tincture put in 1/2 of it again and so proceed as before taught, with the other half you can transmute the base Metallica into SOL, 1 grain tinges 10 shekels heavy into SOL.

Yet a narrow and secret way have the Old Ancients proceeded. After you now have prepared the white flowers, as the COLUMBA DIANAE, so have they caused them to ascend from the calcined FAECIBUS, till it has become in some degree auriferous.

Then they have calcined the RESIDUUM strongly, and extracted the SALT with distilled WATER, purified and crystallized. Such SALT they have rubbed small, and rubbed 4 half ounces SALT to 1 half ounce of fine beaten Ophiris SOL, and melted them, so have they got a blood red SOL, of this they have taken 4 parts, and rubbed therein 1 part of the COLUMBA DIANAE, and put this into such a vessel. (Figure 17).

Figure 17.

and such well preserved, put into our oven in sand,
and let stand so long, till it was incorporated.
Then they have taken it out again, and rubbed 1 part
of the COLUMBA DIANAE, and put again into the double
vessel, placed in sand, and then proceeded, till the
7th. time.

This Tincture they have fused with Ophiris SOL, of
this they reserved half, the other half is mixed
with 1 part of the COLUMBA DIANAE, and thus repeated
7 times. This Tincture is more penetrating than the
one before.

The nature is most hidden, for the dear Ancients
from illumination of the great JEHOVAH saw yet

further, and prepared their Tincture in such a manner.

Thus they manufactured their white flowers, and so do they have all the Liquorem, which they changed, preserved and collected; The white Flowers have they purified, by simple progressing, till they appeared like a diamond; such have they raised, or with the flowing PHYTHON united, and again sometimes ascended with each other, then they have raised up this poisonous reptile and Dragon.

The LIQUOREM they have sometimes distilled over, till it has become like tears from the eyes. There will however not be much of this.

This they have taken, and from the remaining Faecubus, remaining behind, extracted the PIRTRE, and always distilled off their SPIRIT again from the PIRTRE, and this so often poured on fresh MATERIAM, till the SPIRIT has become small; that they could have extracted nothing more; then have they extracted all the Blood, and so always in the distilling off it remaining, which done together, is also raised up. Which is nobler than SOL.

This is called the Blood of the Dragon. The remaining Faeces have they strongly calcined, and extracted from this a SALT, which they have clarified, till it has become pure and like a Diamond, for one must clean it well. Then have they rubbed this SALT small, and put the Blood of the Ancients thereon, and melted such in a vessel of Acures together gently, so has the SALT become blood red, and the red and white Flowers they have prepared again. This they have associated in a vessel for a short time, and got a penetrating Tincture therefrom, also they have taken of the red Flower, or of the Red Lion 2 parts, of the GLUTEN or Columba Dianae 1 part, rubbed such amongst each other, put in a vessel of good burnt EARTH, and with ACURES covered. Such have they added a lid to it, that nothing can fall in, and place the vessel, that in preparing the FIRE, that it is heated gradually, and heat it for 4 hours, so the red flower took the white into itself, and this repeated till the 7th. time.

This is that, which the Ancients said, how they finished their stone in 4 hours. Then have they taken of such 1 part and added it to 4 parts of Ophiris SOL in flux, so will it become pure Tincture; of which they have incorporated half of this with 1 part of the Columba Dianae, as has been taught till the 7th. time, and in such a manner increased their work in INFINITUM, whereby they supported themselves in their need, and come to the help of their poor imprisoned brethren.

For Jesse's son, David; 1st. Chron. Chap. 23, V. 14: Has learned it by preserving the secret arcana of his forefathers, and Moses' miracles. And how could David assist the necessitous, if he knew no way to help himself?

For dear brethren, it is said, he had in his poverty 100,000 talents of SOL and 1,000,000 talents of LUNA.

Yes David, before he could prepare SOL & LUNA, he got bread for himself, for he was not sure.

Of Soloman it is said in Kings 3, Chap. 10, that he made as much SOL & LUNA as there are stones. It stands that it is true, that he had it brought in ships from OPHIR, where however did David get it hither? And why has Roboams son been obliged after his death to live in such poverty? Was he not able to send ships to OPHIR, in order to fetch SOL.

Certainly this King, because he wanted Wisdom, that comes from the Lord, would not have been able to fill his covetousness. But Ophiris SOL is something quite different. Therefore have also many of our fathers, who were in need and care, found such noble treasures that they would help themselves anew. For when they were with God, so was that also with them, as his chosen people.

That he has however denied us his blessing, which is caused by us and our fathers sins, which he avenges unto the 1000th. generation. Therefore call and cry that he may hear us, and convert by repentance. That we follow his voice, and depart from the unrighteous ways. Did we walk in his precepts, then would we also receive prosperity and be glad in God.

Plate III.
After the Separation you must again unite them.

Number 4

Dear brethren, Ah! that you might understand me, what I at present draw before you! For I speak even with you, as with children, not in dark words. Get learning and then set out to do the work, that your poor brethren might be releaved from anxiety. Think yet how their minds are troubled under their oppression, that they cannot once raise their heads on high for distress; therefore my heart breaks, to show you a way, to help you and them. Can you not work, then pray to the Lord, that he will give you wisdom: Daniel Ch. 5, V. 21 & 22. It is forsooth easy and insignificant, if you only loosen yourself from all expensive things in the world.

Therefore dear Brethren, have I here again drawn you a figure, where you may have two ways before your eyes.

First you see, how the Old ALBAON chops off with his sword the intractable Winged feet of the PHYTHONI, and that the PHYTON has in his hands a staff, the meaning of this is, that when such has happened, he then has a two-fold Nature, which the two serpents point out.

This Old Procreator is drawn out once from the PROMORDIALIS CHAOS, that is signified from the FIRE appearing Dragon. But however the Old One floats in the Air, denotes the SPIRITUS UNIVERSALEM or PHYTON, the beginning of all things, as I, dear brethren, have taught you in the beginning. The Old One however has a Scythe or reaping hook, and will cut off the feet of the PHYTON, which is an indication that such also can be prepared from another MATERIA, than out of the ALBAON, which is a black grey and heavy MATERIA, and one can get in abundance. And that the Old One floats in the Air has the meaning, that from this Old One, as I taught you before you shall prepare the COLUMBA DIANAE, and such blend with the PHYTHONE, and once more cause it to rise, and then you get a twofold PHYTON, or the right GLUTEN, the fiery Dragon in the wet way.

But there is prepared from the GREEN LION a crawling Dragon, and you have it before your eyes, and can compare with the Old One the wet way, and the Dragons the dry way.

In the work you go on, beyond, in the dry way in the preparatory work, there is another Modus than in the Wet. Dear brethren, that I may show you, however, that the Ancients possessed more secrets, for they are not all united in their ideas.

Those who had not the understanding, that they could fabricate the LAPIS ALBAOHEST or the PHYTON of the Ancients, yet have they known the MATERIAM, and have been in the position to go to work on such.

They have seen, that the ALATRON could be extracted from the EXCREMENTS of animals, so did they come on the thought that in Man, the most eminent production; in and out of which such excrement was frequently to be got. They have therefore collected from healthy males their Urine, and let such stand in casks to putrefy. Then have they prepared through many distillations a volatile SPIRIT PHYTONIS, as well in a dry form as also a Liquor in an horrible fiery form and Property therefrom.

This they have well preserved, and well kept it in an ALINGEL, then they have here conveyed and have prepared from the Noble Wine their Heaven and SPIRITUS PHYTONIS.

These two fiery Dragons they have poured together, and again driven over, till the most of which ascended into a white GLUTEN; which they have put into an ALINGEL, and raised, then they have taken of the KALII and of the SALT EARTH equal parts, and out of both have driven with strong FIRE, out of a

crooked vessel a strong WATER, or have from these
two stuffs taken only one such mixed with an EARTH,
and with strong FIRE prepared one WATER therefrom.

This they have driven over several times, that no
moisture was there. Then have they put the above
volatile salty Liquor 1 part into a pot, as here
depicted. (Figure 18).

Figure 18.

Such covered with a head with a long beak, place in
a mild steamy heat or B.M. placed in WATER, then
they have prepared the others from the VITRIOL and
NITRE; and also by degrees poured 1 part thereto,
then have these two fiery Dragons bit each other
powerfully, and fought, till they at last remained
lying dead. Then they have put a vessel with a spoon
to the beak, and have driven a PHYTONIS spirit

thereover. When the half also had flown over, they then have taken off the Vessel, the remainder they have thrown away, the SPIRIT which flew up, they rectified once, and have gotten a mysterious Liquor, with which they have gone to work. Others have taken the above fiery salty Liquor, of which 1 part, and added of WATER GAZA the half, and then driven it over, and have also successfully proceeded, and have then also got the SPIRIT PHYTONIS, as a key to the Art, then have they taken our MATTER, it is BISMUTH or COBALT, and beat it small, put it into a vessel; Of this secret Liquor poured thereon, to the height of 2 diagonal hands over the matter, and place this in a mild heat, continued till the Liquor is coloured grass green, the tinged Liquor is poured off, and fresh liquor is poured on, so have they got a grass green Liquor and again a green Solution; these tinged Liquors they have put together into a vessel, and driven it over to the half, the remainder have they put in a cool place, so have the stones crystallized, as Smaragdine.

These they have taken out, the Liquor again drawn off to the half. Placed the vessel in a cool place, then are yet more crystals crystallized.

Then have they made this Secret SALT dry, and placed in a crooked vessel, and laid this in an Oven, that the FIRE should beat around it, and then drive out with strong FIRE a volatile SPIRIT with an OIL, so has the SPIRIT gone over in a horrible Wind. For that reason they have luted a great ball in front of the crooked necked vessel; the ball have they taken away, and there remained behind for them a ALINNEGRA; then they have poured the Liquor back upon the matter, and replaced the ball, and driven over the Liquor again; and this they have repeated 9 or 10 times.

Then they have taken off the ball with the Liquor, and separated into 2 parts, and purified it well, then they have obtained a white and high yellow SPIRIT; with the white they have washed the ALINNEGRA, i.e. they have rubbed this small, and put it into an Alingel, and poured the white SPIRIT thereon, and well mixed them together, placed in a mild heat, and then the SPIRITUS PHYTONIS becomes a blood red colour; this they have poured off, and poured on the above mentioned high yellow Liquor; and the other on the LATONEM, and so long washed, till all redness had been drawn out of it; and that which remained behind was beautiful white, this they have well preserved. They have poured together all the red Liquor, and the white SPIRIT in a high vessel was driven off, till a thick blood come over. This they have put aside, then they have caused yet

more to fly, from the flown over Bird, that all
watery moisture was separated therefrom.

Then have they calcined the remaining EARTH, and
with the Phlegma drawn out a SALT, which they
purified, and then they rubbed therein the Blood of
the Ancients (which is more precious than Ophiris
SOL) placed into a crooked necked vessel, and poured
the flying Bird thereon, and then distilled and
cohobated.

Then have they called such a Liquor their Chaos.
This have they put into an ACURES well secured. The
remaining white EARTH have they taken, whereof they
in the beginning extracted the white SALT, and
rubbed it small, and put it into a vessel of ACURES.
(Figure 19).

Figure 19.

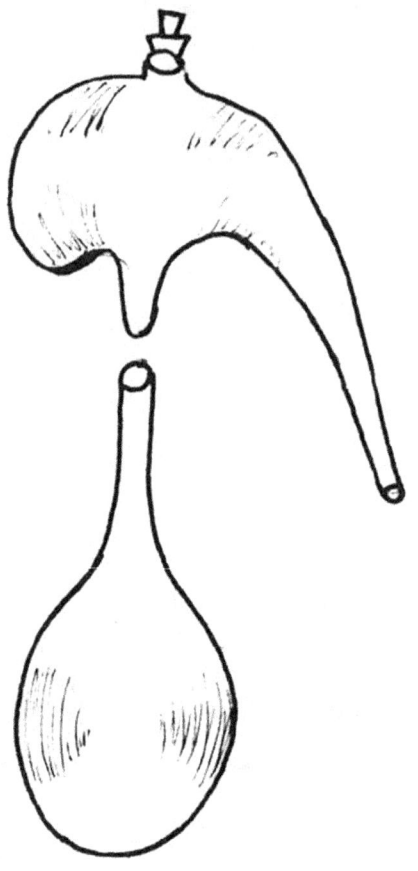

And weighed the same, and added to it 2 parts of the
above Liquor, and then put the vessel in a mild
heat, and distilled over the Liquor, and this they
have COHOBATED so often, until it remains lying with
the EARTH and is fixed. Then they have poured
thereon fresh liquor, and so continued, till all the
EARTH with the Liquor has flown up, and then fixed
in the FIRE and remain lying.

Then have they taken the half of this and melted with equal weight of Ophiris SOL, so have they gotten a clear Tincture.

Then are such carried off, as I taught you before, already dear brethren, because all the after work is one and the same.

But as you see the preliminary works, as you see, are different.

Dear brethren, there have been some of the Ancients who worked thus: they have taken this green Lion, and shut it up in a ALINGEL, i.e. a long necked vessel, and put it in horse-dung, or otherwise in a mild vaporous heat, that it should be resolved. Then have they gone to work with it, as dear brethren, I taught you in the preceeding, taken from the Ancients description of the Work.

They have in the work, proceeded with this wonderful SALT, as the wise did with the Green Lion, so they prepared it from the Old One.

For in the wet way it is a work, which concerns the after-work, further, dear brethren, you should also know that with the prepared Liquor they attacked the PHYTONI; because they had not all the knowledge of this our marvellous MATERIE, and also they have taken the PHYTON, and put it into a high vessel, and poured thereon of the flying serpent, and put a head onto the vessel, placed the vessel in sand, and so has the flying (dragon) eaten the creeping reptile, and disintegrated it. Then have they driven it over, because it has remained from them a useless EARTH, this flying Bird they have then poured on a fresh PHYTON, and again driven them over together, so have they got a secret dissolving WATER, of great effect. This have they let mount yet per se in a well secured vessel, that all remaining watery moisture should be separated, and so get its strength, fiery and clear, as a tear from the eye. With this Liquor have they also been able to do wonders, and have given it divers names.

With such Liquor have they also gone to work: They have taken Ophiris SOL, and beat it into thin leaves, and put it into an ALINGEL and poured so much of the Water thereon, as they thought to be enough.

Then have they placed this in a mild heat, so has this fiery SPIRIT consumed the SOL, and disintegrated it into Blood.

There has the King lain in his Blood, and his limbs sunk to the bottom. Then have they poured off this Red Liquor, and from their SPIRITUAL Liquor of COLUMBA DIANAE poured yet more thereon till the Blood was decocted. Then have they poured such into a high glass. (Figure 20).

Then they put a head thereon, and LENTE distilled over the flying reptile, till it became a dry Liquor. This have they put in an ALINGEL well secured, some have put this Liquor into a crooked Vessel: (thus: Figure 21).

Figure 20.

Figure 21.

And put it into an Oven in sand, and distill this
over and Cohobate it, and so have they got the true
OIL of SOL and so have they received a great
MYSTERIUM. This red fiery Lion have they divided
into two portions, the one half they have put into a
high ALINGEL, and such hermetically closed, and put
the ALINGEL into an Oven; (Figure 22) in which they
could give ALGIR FIRE, and let it stand, till it was
fixed and firm. Then they took it out, and poured of
the other portion, 1 portion thereon, and closed the
vessel again, and put again into the oven, and let
it stand until it was again firm. Then they took it
out, and poured the remainder thereon, and this they
continued till the remainder was all firm and
together, so they found a fiery firm OIL.

Then they caused to flow 4 shekels of Ophiris SOL,
and poured 1 shekel heavy of this OIL thereon, and

thus changed the OIL and the SOL into a Red
Tincture, like a Burnt Blood in appearance.

Figure 22.

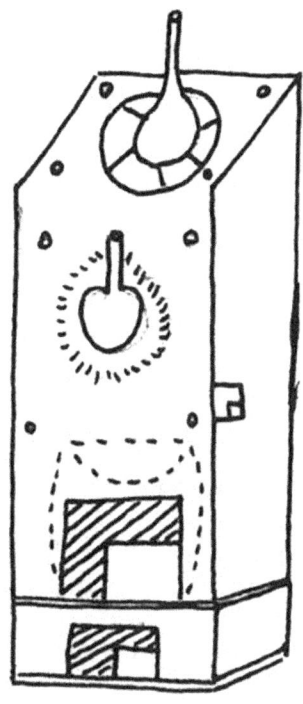

Of this they took a single grain of this Tincture to
16 shekels heavy of other metals, and so they got
the most beautiful SOL.

Dear brethren, others who from God the Lord endowed
with more wisdom, took their PHYTON which now had
wings, and shut it up in the SOL; the SOLUTION they
poured off, for their COLUMBA DIANAE was nothing
strange to the SOL, and then weighed it, and if
together 8 shekels heavy, so they placed thereto

also the common PHYTON, which must be however pure. This they put into an ALINGEL well closed in the Oven in a mild heat; in the beginning it was black and a true Chaos, and went at last through all the colours. Others, who saw yet a nearer way, poured the COLUMBA DIANAE on pure LUNA, and dissolved it in such. This heavy WATER they lifted up.

Then they took Ophiris SOL, and dissolved it also in one part of this LIQUOR, and poured it off, also into a clean vessel; then they took of the GLUTEN or of the COLUMBA DIANAE 1 portion, and poured it into an ALINGEL, and also poured the Blood of the King 1 portion, and closed the vessel, and placed it into an Oven in mild heat, and let it stand so long, till it began to become black. Then they continued the ALGIR FIRE till all had coagulated together, then they took it out, and rubbed it small, and put it again into an ALINGEL and poured of the COLUMBA DIANAE 1 portion, and also of the Blood of the King 1 portion thereto, and closed the vessel, and put it again down and let it go through the colours, and this they repeated 7 times. Then they took this Stone out of the vessel, and let 16 shekels heavy of Ophiris SOL flow, and added 4 shekels of this fiery Lion thereto, and so have they got a Tincture.

Of such they have only thrown some grains on a whole tit of other metals, so have they got the most beautiful SOL.

Afterwards now the dear Ancients did in such a way, help themselves many times in need, as also their brethren, but how would they have wished to remain in such need, if the Highest had not helped?

Therefore, dear brethren, have I disclosed to you anew a great secret, that you might find a consolation for one cannot have everywhere, what one desires in time of need. That it may not fail you in your grief and misery, till the deliverance comes according to the words of the Lord.

DENI, ADONAI, BOCITTO, OCHYSCHE in quick time. Persevere and guard yourselves that you do not unite with the heathen.

Be not disconsolate without ELIAS and Our King, to collect again the heathen to you, to destroy them, as some of our brethren did, with whom need made them bad, that of them often 1000 have fallen in one day as by BARCHOCETA. This one thought because he also understood this Science, how one should prepare the MYSTERIUM, and to the people to prepare a dreadful poisonous WATER; thus, you must take PIRTRE

KALY PHYTON PUVON SALT EARTH, out of this he taught the people to make a poisonous WATER, and they made by Cohobating such a horrible fume of poison thereby. This they poured into Springs, and it was thick, dark and muddy in the heaven. They poured the same into a vessel, they put the same in profusion to the FIRE, so it began to smoke, and poisoned also the AIR, that men and beasts perished, then there came such a frightful illness among the people, that they got dreadful burning blisters, which then began to putrefy and to stink, many became pitch black, and fell suddenly down. If it came into a house, the poison raged so strongly, that no escape was at hand. Guard yourself, I say yet again, guard yourself from the like, that you do not make the burden heavier among you. Would you walk in the commandment of your God, so will he guard you, and send a speedy deliverance. Rely upon Him who made heaven, and earth. Hope and wait patience.

Guard yourself from errors. Remain faithful then your work will soon get its reward. Help comes out of Zion. Psalms 53.

Plate IV.

With this cut off the feet of Phython or burn them off by Fire prepared from the Green Dragon.

Number 5

Dear brethren, our father Jacob served 7 years for Rachel, and it was very toilsome for him; and his reward was changed for all that, considering that instead of Rachel he got Leah.

This is even truly a prototype of your servitude.

The Lord, the Holy One however will soon conduct us home, when the other seven years of affliction are past.

We must suffer with Leah the weeks of affliction yet a short time, then will Rachel be brought to us.

Consider that Ephraim is the first born of Rachel and not Judah. Therefore are we forsaken, and the Lord calls Jeremiah, V. 30:20. O Ephraim, my only son how my heart breaks for thee that I must again have compassion on thee, and deliver thee and lead thee out from Hagar and her son.

Brethren, the time is soon at an end, that Hagar will drop her son for sadness. For the holy ADONAI will thrust her out from our inheritance, and she will languish from heat, and not be able to see, that her son dies from sadness, till the Angel will come of the Chosen People, to comfort her a little; yes the Lord will not destroy all the heathen, but preserve a portion for the service of his people.

Dear brethren, to give you yet a consolation till then, and to cheer you, have I again here drawn a figure in order to impress on you right deeply the secrets. You see an Old hollow Oak-tree standing in a garden around which is twisted a rose bush with red roses, which has gilt leaves. Underneath from the stem of the tree runs a LUNAR white WATER.

There are some who not far from this have hoed and digged, but found nothing; except those who by the way, who contemplated the weight. Dear brethren the old tree is our black and heavy rocky lumps of our ALBAON.

You must strike this rock, till it gives WATER, as you have already previously with all circumstances been instructed and taught. For the LUNAR white WATER, which flows underneath from the tree, is our PHYTON COLUMBA DIANAE, which is heavy. That however the tree bears red roses with gilt leaves, such signified the BLOOD of the Old One, which must be drawn out of such visibly, as the BLOOD of the Lion, or our secret.

That it however flows underneath out of the tree
signifies a LUNAR white WATER, which should be
prepared from the root of such a thing, so that in
all parts of its nature it shall be KIN to the SOL
and LUNA, and that it may be also easily made firm
and stable.

If you should lose all writings, then you should
only depict those Figures, or draw such for your
children, so will they, as others who are of a good
understanding, easily understand such, for one dares
only to speak with few words, what concerns the
MYSTERIA, come to the help of it.

Dear brethren, in order that I however suppress
nothing which could serve for your advantage, so
mark what wonders the Ancients have accomplished,
when they have taken the PHYTONIS or our GLUTEN
AQUILAE, that you might yet better understand, our
COLUMBAM DIANAE, and with such have also gone to
work.

They have taken this as it ascended from the Old
One; and have caused it to fly over the mountain,
i.e. over the ALEMBIC, for 7 times, so has it become
brilliant, but in thus doing it the Basilis Kischer
way. (Figure 23).

Figure 23.

Then have they prepared an acrid WATER, as you well know, and is well known already to you from the foregoing; in such have they dissolved LUNA, and taken such a SOLUTION, beat again with NITRE WATER, and well sweetened. Of such have they rubbed 1 portion among 2 portions of the COLUMBA DIANAE, and then put on the fire, but before they were placed with each other in a tall vessel, therein in a most tolerable FIRE the LUNA also ascended; this they have done among the remainder, and again the Bird (Figure 24) caused to fly, and that to the third time; so has it appeared with a great brilliance, and has also tolerably lost its poison.

Figure 24.

The remainder in the crooked necked vessel have they taken out, or put into another, namely, what remained in the LUNA, and raised it up. (Figure 25).

Figure 25.

The COLUMBA DIANAE however from the head put thereto, and thus with each other metamorphosed into a heavy glittering WATER. Had they however poured this WATER several times on the remaining LUNAR EARTH, and thereof caused it to fly, so would they

have it coagulated into a White Stone. For this
serpent or three-headed bird loses at last it wings,
and will remain firm, such has happened to me, dear
brethren.

Figure 26.

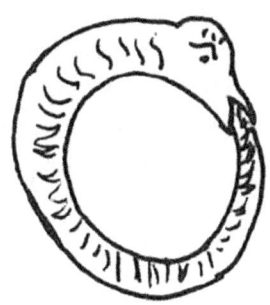

This glittering stone metamorphoses the VENUS into
good LUNA, 3 to 14 grains to 16 shekels heavy of
VENUS in flux, and this stone can easily be
augmented, as you will have understood in my
previous writings. This is the METAMORPHOSED VENUS
cloathed with the LUNA.

Also has it happened to me sometimes that I poured
the COLUMBA DIANAE to a Ophiris SOL, brought before
to a DUST, and driven it with such aloft, in order
that this green BIRD might rob the KING of BODY and
BLOOD, and raise it also.

As I diverted myself also, and that such should fly 7 times, and so it remained behind, and even incorporated with the Queen, for that reason.

Figure 26.

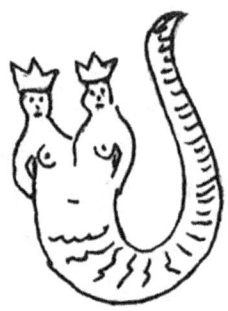

I was constrained in order to see what would here end. But they were firm, and their wings quite burnt off and altered itself into a brown-red stone, i.e. the King became fixed with the Queen, and both became in a short space of time homogeneous.

This brown red Stone tinged and metamorphosed, yet for all that like the above with a few grains some shekels of metals into SOL.

Therefore when you will in a short and little time help yourself and your brethren, then let this be said to you, that all Metals ex PHYTHONE have their source and beginning. All Metals consist of EX-PHYTHONE, for the MERCURY is their beginning. This

sleeping Lion you can with the fixed DIANAE easily
again waken, (Figure 27)

Figure 27.

if you rub only of the COLUMBA DIANAE 1 portion
among 4 portions of the fixed, and put such only
into a common vessel, then will the fixed one
overcome the flying, and eat it in a very short
time, so that you receive here a higher arid more
penetrating stone and can likewise augment as long
as you wish. (Figure 28).

Dear Brethren, that you can bring forward no excuse,
as if the wise Creator did not care for you, so have
I from inclination and command recorded, and must
record, that it might not get lost, and thereby you
might obtain a consolation. For it is easy,
certainly not to all men; for many have such a weak
understanding, that it is

Figure 28.

impossible to comprehend it. To some others however
it is only Child's play. What do you suppose indeed?
Should Moses and also his brother Aaron not
understand the same?

Yes! For how could he turn the Golden Calf into
ashes? Exodus 32.

The Spirit of the Lord was on him, for he even saw
before, how the holy ADONAI created heaven and
earth, and from what, and how this efficient Spirit
yet hovers before his eyes. This intelligible Spirit
now, which was the life of all things, Moses took
and consumed by fire the Calf, and made it into

powder. Also Miriam, Mose's sister, because she was leprous, was cleansed by the help of this secret.

And this was to the Ancients in their need the greatest consolation. This secret was, it is true, also known to the heathen through carelessness, but it has vanished again from their hands. Therefore I ask you that you do not slight my warning and such, and where and how you can conceal it, that it be not lost.

It is in Roman, but the most is described in the Arabian tongue, that it be not read and understood by everyone.

Dear brethren, the Ancients have at their sacrifices often had no FIRE necessary, for such has been lighted at command of the most High through the Angel Michael. So have the Ancients also had a WATER; 1 Kings, Chapter 18; which they poured on the sacrifices, then such has taken fire.

Such WATER have also some had in a dry form. Now this is true of ELIAS, when he will come with the Messiah, Gods' and David's son be found again, for it lies yet concealed at Jerusalem, in the Holy City, as I said in the beginning. That you may know however, how this was prepared, and can be prepared, now observe.

You must understand Nature, otherwise, it will be to you incomprehensible.

One finds a SALT, which then burns, it is the NITRE, this must be cleansed by means of dissolving and coagulating, it must twice be precipitated from a SOLUTION of Spiritus Vini Rectified, in order that all incombustible mucus be taken therefrom; N.B. which in all its parts, the truest, yes a pure FIRE. It is prepared from the EARTH, and is to be found everywhere.

It is not the Universal SALT of the EARTH, but it is quite another; yet it has much in common with the same; the most of it is found in the earth, where animals as sheep & etc., have their standing place. From such a SALT have they driven out with great care a fiery red SPIRIT, this they have poured on PYRTRE, which is pure and clear, and so long COHOBATED and distilled, by means of a crooked necked vessel, till the PYRTRE has become a WATER with it.

Figure 29.

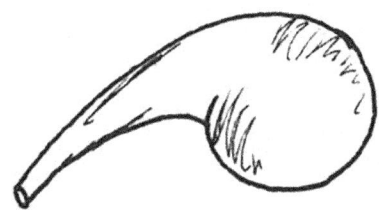

This have they poured again on such a fresh SALT,
and so it has become a thick liquor. This they have
once driven over, and again poured it back, so that
it has become a thick Liquor. When this stands in
the cold, it will coagulate. Of this they have taken
1 portion, and put thereto 1 portion of NAPHTA, in a
vessel of burnt stone, for it grows very hot, and so
it is prepared, and on everything that it is poured
on, is eaten and consumed by fire, for no WATER can
exterminate it.

For that reason take care, that nothing of this
preparation comes in contact with yourself, for it
consumes flesh and bone. Should however some of this
touch you, then take only EARTH mixed with SALT, and
moisten this, that it becomes like a jelly and apply
this, so it will be put out, otherwise nothing in
the World can extinguish this. The NAPHTA however is
an OIL, such as flows from the Rocks, and is thick,
when it comes to the AIR.

Also you have now information of the ש א that your children can or may find it. What however concerns the said OIL, will be found at the time, when the Lord collects us again in our Land. For our fathers, priests and Levites have for such a time of need prepared. For no sanctuary, so we also require no holy consecrated OIL, for the finery is taken from ISRAEL. Yet that you might know, what this OIL is for, and whereof it is prepared, so mark:

The Ancients took a vessel of ACURES or a vessel of stone, covered with ACURES, and placed this on a FIRE (coal fire which did not scorch or give flames) and put 32 shekel heavy of good Olive oil therein, when this became hot, then they had on hand tender, bruised and rubbed MYRTHEN, CINAMONEN, CALMUS, CASSIA, of each 8 shekels heavy, this they stirred amongst each other, and let it stand well muffled up, that the strength may remain together for 1/4 of an hour.

And then they took this from the FIRE and allowed it to cool, and then squeezed the BALSAM through a new virgin cloth and preserved it in a clean vessel.

Herewith were the Kings and Priests anointed.

Guard yourself however, that none understand how to make this ointment; he must be a Priest or from a priestly caste. Exodus 30. God the Lord has forbidden it, so none dare to fabricate this holy ointment OIL of the bridegroom, but only a priest, who used it at marriages, when two betrothed persons appeared to the Priest, so was the bridegroom anointed with this OIL together with the blessing of Jacob.

All our brethren have also received the anointing that their matrimonial state may be blessed, and they become fruitful, for we are all children of the HOLY ONE.

This ointment Oil was prepared out of 32 shekels of Olive Oil, 12 shekels heavy MYRRH, of ALOE and CASSIA, 8 shekels heavy, Exodus 20. The preparation is like the previous. This then the Priests preserved also in a pure and holy place. With this they anointed also the sick and blessed them, so they became often better or died. Were they not annointed before death, then the Priest, after death annointed them as well on forehead as on breast. Dear brethren, out of what the perfume consisted, you shall also know.

The Ancients have also put it together. They have taken STACTE, ONYCHA, GALBANUM and FRANKINCENSE **a͞a͞** and such also preserved. Exodus 30.

With this have the Priests in the Holy of Holies strongly been obliged to incense.

This is now what belonged to the Priesthood, that they could show such to their children. What however concerns the URIM of the High Priest Exodus 28; in which the great JEHOVAH showed himself, and through such at certain times spoke with the people, has not been lost, as one says, but it lies with the whole priestly finery preserved as I have previously announced to you, and will at the time, when the Lord again will visit his people, be brought out by Elias.

How my soul rejoices when I think thereon, that at that time all will be renewed and there will be no more injustice. Israel will rejoice, that the 2300 days are over from evening till morning, Daniel 8.

Be comforted, and abide in patience, and it will to you be richly rewarded.

69.

Plate V.

So you get from our old Oak tree (OLD ONE) the White and Red, make therefrom a water and observe the Weights.

Number 6

Dear brethren, my heart will burst, when I see your distress. Shall they rage and then ruin without cessation, and the sword devour the mother with the children? Shall then innocence count for nothing? That you even forget us.

Lord, will thy anger not become tired to destroy? Is it possible that the Covenent of Abraham, Isaac and Jacob shall be no more?

Our fathers have acted unwittingly, and have disappeared, because they smote the shepherd, Zachariah, Chapter 13.

Our fathers misdeeds are certainly great, and we are also not without sine They have persecuted the righteous and killed them, who showed thy great name.

They have passed, and have with other gods made idolatry. If you sent them a Seer, so would he be destroyed by them, as if they were such as agitated the people. Therewith they laid their hands on the Holy Ones, and those who were sent by Thee. This blood cries to Thee from the earth for revenge. But ISRAEL, thou art struck with blindness till the Champion comes, and will again put thee into thy inheritance. How willingly, dear brethren, would I help you in your need, that is known to the Lord, therefore can I not forbear to discover to you yet something more. I have again here depicted a Figure, and how can I otherwise, that you may yet see.

You see a flower with 7 leaves, which flower is red and SOL, the leaves however not SOL, which are blown by the North Wind. The leaves signify our GREEN LION, which is far better than Ophiris SOL. The Flowers however signify our RED FIXED LION, the TINCTURE, which no north wind can move. And that it stands on the mountain, has a two-fold meaning. Firstly, it is sought by many men, but found by few. For this Mountain is our Altar, our MATERIE, which is watched by nothing but griffins and dragons, i.e. they are poisonous in their first preparation, therefore are they feared, since yet their poison is pure TINCTURE.

Secondly, that such Flower is born in the Air signifies that it mounts on the mountains, i.e. into the ALEMBIC.

N.B. That is the COLUMBA DIANAE and is at first all there is found and seen. That however many griffins watch the mountain, and fiery dragons, has a two-fold meaning.

The fiery dragons are the PHYTON or unveiled DIANAE.
(Figure 30).

Figure 30.

That nobody knows her, as well in the wet as the dry
way. So can also with the help of the UNIVERSAL
CHAOS of the Ancients, like from the SALT AERUM of
the Magnetic fatness of the AIR this fiery Dragon be
understood, with which the Ancients did wonders, and
this they have learned from Moses.

The Griffins are, as previously mentioned, nothing
other than SPIRITUS PHYTONIS, which can be prepared
as well from the old Albaon, as also from the CHAOS
UNIVERSALI; from men and Vines. (Figure 31). With
which they mixed some MINERAL spirits, and from such
they prepared a volatile SPIRIT, with which they
burnt off the wings of the PHYTON, and have

metamorphosed into a viscous WATER. Now however, dear brethren, shall I show yet some secrets worthy

Figure 31.

of wonder from our inexhaustible spring of the Ancients, and yet more Mysteria. When you, as you previously understood, your PHYTON, or Flower have cleansed by many ascendings, so have you put this among prepared LUNA, or transported into, and have also with each other let fly once, so has it become fat and glutinous.

This you have preserved, and when you have prepared it, you will likewise see its form. This mercurial Flower will appear, like the most beautiful pearls. Will you prepare large pearls, then take only small ones, and make such to an inconceiveable DUST, and take of the COLUMBA DIANAE, so much thereof, that it becomes like a pap, somewhat thick, and has SOL or LUNA forms, how great you would have these, made into such pearls, and stick through each a strong bristle, which is clean. The bristles stick to a cross piece of wood, this put into a vessel, which you can well close; and place it in a mild heat, (Figure 32) so they will become in a short time as hard, as they have been, and of great brilliancy. Take these out and steep them into the Liquor prepared from the COLUMBA DIANAE, arid let them be wiped clean, then put them again into a glass, and in a mild heat, the others in a short time will surpass. You can also cause forms to be made with divers figures, as birds, lambkins, pears, apples and the like, besides that, you can prepare pearls of priceless value.

Figure 32.

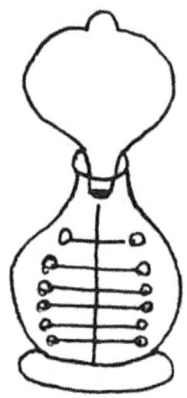

For one is not everywhere in the position SOL and
LUNA to expose to sight, for the enemies sake,
because, if they became conscious of it, they would
torment us badly. For consolation in your misfortune
have I disclosed this.

One can carry such a pearl himself, and conceal and
yet be of great value. The Ancients have yet further
looked about in Nature. If they had precious stones,
which had not got their mature and right colour so
have they quickly known, such to bring to sight. If
they have had a Diamond, which was large and not
pure, they have cleansed it from all dirt, and put
it into a vessel, and poured thereupon the COLUMBA
DIANAE, and let them stand with each other in a mild
heat, and then has it received its pureness and
beauty. Others have, however, taken the COLUMBA

DIANAE in dry form, rubbed gently, and put such into a vessel of good burnt EARTH, (Figure 33) and laid in the Diamond, and put thereto so much of the COLUMBA DIANAE that it is only covered. The vessel they have well closed, and then put it into the FIRE, and let it stand some time, then have they opened the vessel, and found their Diamond surrounded with a skin. This they have separated from it, then they have found it larger and prettier. Yes they have taken only one right beautiful Crystal, and with such in this manner proceeded, this they have now performed in all secrecy.

Figure 33.

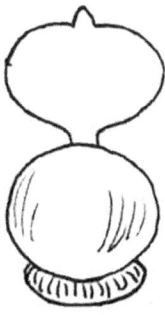

Others have taken RUBYS, because one can have these in considerable size, and have gone such a way, and have prepared DIAMONDS thereoutof, which shone like a lightning flash.

They have therefore not alone in such a way maintained their life, but have also come to the help of their poor brethren, for this have they easily concealed.

Dear brethren, how often have you with such things by permission of the great King of the World appeased your enemies, that they have become compassionate, and have discharged you from servitude. But we must complain. We are everywhere obliged to serve, our inheritance lies in waste, and has become a portion to the impure, our houses to the enemies, our mothers are widows and we are dispersed orphans without fathers. Ah! Our fathers have done wrong, and merited such with their sins, we must bear the misdeeds from child to child's child. But Lord! What have these sheep done to Thee; these miserable Ones? Hear yet the cry, and the sob of the same. O pity us, that we also lie with our fathers in the same condemnation, for the crown of our head is gone. But thou great King of the World, who remainest eternally, send yet our salvation and our Deliverer, who will lead us out.

Dear brethren be not tired to cry and to call till the Champion comes, who will deliver Israel. For his heart will break, that he may come to the help of the troubled, and to the mournful for joy, to the hungry for food, and to the thirsty for drink will be. Be not impatient, that you may make no more of mistakes to the Lord, and it again may happen to you like to the fathers in the wilderness, and die among the heathen. Fear the name of the Lord, then will rise up the Sun of Righteousness.

Plate VI.

Let it climb to the top of the Mountains; then drive
them together over (the Mountains). So will the
fiery creeping worm be prepared or the winged
Griffin.

Number 7

Dear brethren, that it may not fail you in the instruction, so I will here in conclusion, will not hold back from you the remaining secrets of the Ancients, but will reveal these, wherewith you may be able to come to the help of the necessitous. I have again drawn a Figure, which comprehends much in itself. Give heed to it well.

You see, that a King stands in the Figure with a naked sword; his soldiers are killing innocent children, and they collect the blood in a well standing thereby, which this is already filled with blood, in order to colour it yet some more. Into which descends SOL & LUNA to bathe themselves therein. This Figure has a two-fold meaning, as well in the wet and dry way.

That the King stands, and has a sword in his hand, means, that one shall kill the King with it, which means nothing else but one shall take Ophiris SOL, and with such a double edged sword, with the double Central FIRE, burn up and disintegrate, where I have shown you also, two ways, such to get it over.

The first is prepared out of the UNIVERSAL CHAOS, the other from the ANIMAL, Vegetable and MINERAL.

Particularly must you conform to my teaching, where I showed, how the Ancients therewith went to work. They have taken this fiery flying SPIRIT, and poured this on the PHYTON, and so has it taken away from it the metallick nature, and has become a GLUTINOUS white water. With this have they further proceeded, and this LIQUOR they again poured on fresh PHYTON, so has it also became a thick viscous water. Then have they taken the King, and dissolved it therein, then has it become red like to BLOOD. This BLOOD have they now also called the RED LION, this have they fixed, and as I already taught you, augmented their work with this sloppy PHYTON, and also found a blessed ending.

Others have prepared these two FIRE'S in their volatile form, and taken and dissolved the King with this, and distilled therefrom the flying BIRD partly, to a red LIQUOR. This have they preserved, and put to one side, then have they also dissolved of the DIANA, so have they got a blue-green water; of such have they likewise from the Liquor drawn the flying BIRD in a mild heat, whereof they could not hinder, that this SPIRIT, as well from the King, as whose spouse should not have carried and taken something away.

On this account have they poured this flying BIRD with solar and lunar feathers on the PHYTON, so has it again dissolved itself again in such. Then have they poured off the clear Solution, and put again on a fresh PHYTON, so have they got a fat and heavy LIQUOR, then have they the King and Queen allied with each other, and poured together. Of such have they put the half into an ALINGEL, and as heavy as these two weighed, poured thereon of their fatty heavy LIQUOR. Then in such is the beginning and efficacious and have also herewith sealed hermetically their springs, with each other well closed, and on our oven allowed to go through the colours, till such had become firm. Then have they taken it out, and rubbed small, again thereon poured of the BLOOD of the King and Queen 1 portion, as also of this *CADMI* one portion, and however with each other let it become fixt. This have they repeated, till the BLOOD of the King all had been consumed.

Then have they taken out their Tincture, and divided it into two portions, the one they have caused to flow with equal parts of Ophiris SOL, and so has it become a pure TINCTURE. Then have they again divided this Tincture into two portions, the one portion have they kept themselves for sustenance of their life, the other have they however blended with the half of the unmelted Tincture, and rubbed amongst each other, this again put into an ALINGEL, moistened with the *GLUTEN AQUILLAE, that* it had not only become rather thickish, closed the ALINGEL well, and this again let it go through the colours for FIXATION arid this have they practised in INFINITUM.

N.B. That should you know however, that this Tincture cannot be brought so high, as that described in the foregoing, which took its origin from the old ALBAON, or otherwise prepared from a volatile solar Materia or Metallic Seed; for the Well in this Figure from which the blood red water gushes, signifies our BLOOD of the Old ALBAON, which pure volatile Tincture is, in which SOL and LUNA bathe themselves, or in which they grow young again.

This Figure also signifies the inexhaustable augmentation of our great work, and is nothing other than our three beginning parts whereof our single MATERIA take their origin, with and through which is ALL IN ALL, as well under as above the earth were born all MINERALS and METALS, which Materia called by its right name, also the analysis clearly taught. Will you on this account, in all truth remember this much thereof.

Our Old One according to appearance is like a lead ore, but in its parts is a pure volatile SOL and LUNA; it dissolves almost all its Corpus into a WATER by *VITRIOL & NITRE* prepared, which is a wonder that VITRIOL can be metamorphosed; one finds also often that our Old One has white poisonous bones, which is a pure poisonous PHYTON or SEED, and is here a token, that Nature seeks to make this once again volatile, for in such a form it can become no METAL; for when this comes into the open FIRE, it flies all away, and leaves scarcely a little Solar LUNA behind.

Do not be terrified before this poisonous reptile, for this Old One lies invariable and lives, yet it is a living MATERIA, which breathes and exhales without ceasing, this one without ceasing it smells through a whole room, and when our Old One is triturated and brought into the smallest molecule, annexed to such and is put together into a vessel, then this poisonous serpent coalesces again; so indeed, that if it is often driven from one vessel to another, and coalesces so, that one can with no mallet can often bring from one another; and that this is such a metallic PHYTON or pure Seed, is shown by its high dark blue colour, which with help of white sand or Quartz, or when it is mixed with powder, from the stones; (as David employed it, from the brook with which he killed Goliath) so it gives in a strong fire a heavenly dark blue ACURES, that also its Tincture and Strength shows itself in all places.

When this MATERIA is not yet too old in the mountain pits is met with, then it appears as a LUNA with red intermixed, and is called a volatile LUNAR ore.

If this MINERAL is older and has stood longer, then has it all the colours of the world, as LUNAR white with violet or blue, red with golden little sparks intermixed, often quite pure.

The third species is, when this in the pits begins to become white, then the colours are partly lost, and the MATERIA becomes silver grey mixed with much white. There is in this a most poisonous and pure volatile Tincture, which is well to bear in mind, and all three are of one species and from one root, only that one possesses more Tincture, than the other, yet are all three precious in the Art, and this MATERIA is a right HERMAPHRODITE, i.e. of masculine and feminine seed. The other MATERIA is also not to be thrown away, for it is almost in all its parts a pure volatile SOL seed, looks usually like pure SOL, is found also in red and yellow MARCZ, in black and yellow grains, also in grey and white sand mixed with black grains. From such may it also be dissolved with a LIQUOR, and brought into Crystals of strange properties.

I have also here pointed out, what the MATERIA is, which is denoted by so many strange Names and Figures for it is everywhere easy to get. All other Tinctures as I have shown and described to you, from other MATERIA are only helpers in need, and take away often more times than when you had the correct MATERIAM; if however one cannot have this, while one is not qualified in all places, so will I on that account yet further show you, when you have no SOL or LUNA, as from the base METALS a TINCTURE prepare, by means of the above named double fiery spirits.

Others however who had no SOL or LUNA at the beginning took their volatile FIRE and MARS, and made it tenderly into DUST, and poured thereon the LIQUOR, and placed the Vessel in a mild heat, and so it dissolved the MARS into a high green colour.

This Solution they poured off, and poured fresh thereon; this they continued till all was extracted, the like they did also with VENUS, then they poured these two together into a very tall vessel, and allowed to fly a little more than half of this SPIRIT to distill over.

The remainder they placed in a cool place, so they gathered Crystals as Sapphires and Turquoises.

These they took out, the remainder, which was yet joined to no Salt, they took, and allowed the BIRD to fly over again to the half, and placed the remainder again down, then added the remaining SALT afterwards.

Here they had a marvellous SALT, so in its interior a pure BLOOD, and Tincture, for the above fiery WATER takes only the spiritual body of the VENUS and MARS.

The fiery SPIRIT which was drawn over from the VENUS & MARS, they took, and poured it on the PHYTON, and so dissolved it in such.

The solution they brought also into a tall vessel, and distilled the fiery Humidity therefrom; for the PHYTON does not ascend easily, so they met with such a heavy mucilaginous and viscous WATER, and this they preserved well.

Then they took their marvellous NITRE, and rubbed it small. Here they were not united in the work, for

some brought it into a crooked necked vessel.
(Figure 34).

Figure 34.

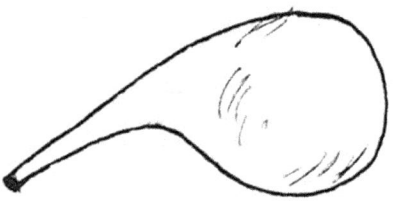

And drove from this a SPIRIT and red OIL with force
of the fire. They cohobated the SPIRIT with the OIL
from the remaining red EARTH so often, till the most
part of it had flown over.

From the remaining EARTH they extracted with the
laid up Humidity, so they distilled off from the
PHYTON a white SALT, this they also brought into
their LIQUOR.

These three now they put in a mild heat, that
they should unite with each other. Then they
distilled such with each other, with a strong FIRE &
COHOBANDO from a strong vessel, then they got a
LIQUOR of marvellous effect.

Of this they took one portion and of the LIQUOR in which the PHYTON had joined, also one portion, these two they poured together, and distilled them afterwards with a strong FIRE COHOBANDO.

They took this LIQUOR and divided it into 4 portions, the one portion they brought into an ALINGEL, and shut it well, and let them together go through the colours. Others however, who were more knowing, took the EARTH, which from the NITRE remained behind, and heated it throughly, and it is well true, and brought it into such a vessel, which was tall, because such was of a virginal essence, and poured on the LIQUOR, when the EARTH weighed one shekel, they poured 7 shekels heavy of the LIQUOR thereon, and put the vessel only in an oven in sand and began to distill, and distilled what would go over. The over ascended they cohobated, till it all remained firm behind. Then they poured on fresh LIQUOR 1 portion thereof, and continued such, till it would no more be coagulated, but flew into the vessel together, fixed and constant in the FIRE.

Of this LIQUOR now carried over, they poured 1 portion to 16 portions of VENUS in flux, and got the most beautiful SOL.

Then they took of this SOL 4 portions, and let it flow, and threw 1 portion of their Tincture thereon, then it became pure Tincture.

This they parted into 2 portions, and put the one portion again in, and worked immediately, and were helped in their need.

Others, however, who were more sage, took this marvellous SALT, and brought such into a vessel, closed it and put it into a mild heat or in horse-dung, in order that it resolve itself into a grass green LIQUOR, which they likewise gave many marvellous names. This they now took, and brought it into a crooked necked vessel of ACURES, and separated this Liquor from each other, and put it again together, as I have already taught you. This now of a simple conception was made with great trouble from this salty LIQUOR, as mentioned previously, with strong FIRE, which alone they made firm, and had no knowledge of the PHYTON, how it as a co-agent, as SOL and LUNA must be in doing this, that they also its seed gave thereto.

When now that they carried their Tincture to the VENUS so it indeed became also a Tincture, when however they carried this to other METALS, before they carried it to the VENUS, as that previously

happened to a SOL containing VENUS. Here they knew
not how to help themselves, and if they carried this
TINCTURE to the PHYTON, so it would not take it in,
for it was yet in its combustible bodies, therefore
they must be satisfied with what they had, that
their work was not once satisfied.

Dear brethren, Nature is most hidden, yet she can
give no more, than she herself has in her power.
Others now who had no SOL or LUNA in readiness, yet
from a mighty understanding when they prepared their
double fiery spirit, they preserved such, and made,
as I taught you already in the beginning, a solar
SPIRIT with ALATRON, HADIT, CELUVIALATEL, this they
dissolved in pure WATER, and let such again
therefrom fume, this they repeated several times,
then they poured a SPIRIT prepared from the VINO
thereon, and extracted all solar Tincture thereout.

This SPIRIT they now distilled again, till there
remained a red DUST, on this DUST they now poured
their fiery WATER, and extracted it several times,
then they got an auriferous PHYTON, then they made
with the fiery WATER also from the PHYTON a
glutinous LIQUOR, and poured these two together, and
made this according to the Art firm; then they got
an OIL TINCTURE on VENUS in SOL.

Others however made a SALT from the MARS and VENUS as is taught and from this with force of the fire drove an OIL LIQUOR and poured these 3 in similar weight together. The EARTH VIRGINEAM so remaining, they calcined with strong FIRE and commenced to carry their Tincture thereon, as is already said, and so they got a Tincture much stronger than the former, and could also easily increase this.

Therefore can you see, dear brethren, when you wish to, that the ALL WISE KING of the World has shown enough ways, to help you in your need. That however I keep nothing back, what can serve only to your well-being, so have some of the Ancients also gone in the following way, for they not always would have been able to attain their purpose, and have yet understood and known the true MINERAL MATERIA.

So have they taken such, and beat small, and with this MENSTRUO or fiery WATER dissolved its body, and prepared from the same a marvellous SALT, and have come much nearer to Nature.

With such a GREEN LION have they gone to work, according to the above teaching, and have also obtained their purpose.

Others who knew not how to help themselves, and who wanted wisdom only merely with this fiery WATER unfolded the Ophiris SOL into a blood red Tincture, and again mildly distilled such a fiery WATER therefrom, till a red DUST remained.

This DUST they have carried to other SOL standing in a red flux, and tinged such SOL with it, and transmigrated it to a brittle red MASSA, with which they have brought some portions of LUNA into SOL. I could, dear brethren, reveal yet more of the same secrets, but the life of man is too short, but I have only wished to show you the most secret and important ways, that you might have something for consolation in your oppression, in order to free your poor imprisoned brethren from the bonds of servitude, and come to the help of poor widows and orphans, and to provide for the necessitous and miserable, that you through doing will eradicate all evil.

Give alms willingly, fast and pray, for such eradicates sin and releases from death. You will support yourself that you may live. Tobit 12.

Hold this book even concealed, that the curse may not come on you, and you be banished from the earth. Act wisely and wait in patience. The blessing of

Abraham, Isaac and Jacob will come to you, and God will fulfill the promise of Abraham.

The Lord will keep his covenant, and he has sworn. Come all, you chosed ones to praise the Lord, the King of all the World. You servants of God praise the Lord for his gifts, and praise the eternal Saviour, who lives forever.

Praise with me and all that is by us, praise the Lord! Praised be He, who releases His people, and raises from the dust.

May His Kingdom remain to Eternity.

HALLELUJA, HOSSHIANNA, HALLELUJA.

FINIS.

Plate VII.

In this Red Child's blood dissolve the King or Queen so will the Sun and Moon themselves bathe in it, for this well is inexhaustible.

A Few Arabic Terms

A Short Explanation of terms from the Arabic used in this Book.

ACANUS - First Metal

Asophol — Gold

Marcz - Clean virgin Mars or Virgin Earth

Acures — Retort

Algir - Fire

Albaon - Plumbum Nigrum (lead), Magnesia, Bismuth

Puck or Puch - Stibium (Antimony)

Phyton — Mercury.

A Word from the Publisher

Thank you for purchasing this small work from The R.A.M.S. Library of Alchemy. During his lifetime, Hans Nintzel was dedicated to the identification, acquisition, study, retyping and, when necessary, translation of what he considered to be the most important known works on Alchemy. Hans was assisted by his sparse network of fellow Alchemists, all members of the Restorers of Alchemical Manuscripts Society (R.A.M.S.). I was an active member of R.A.M.S.

My goal is to publish all of the works originally made available through R.A.M.S. as photocopies. To facilitate this, I have chosen to have the books professionally printed. I also have a few titles that I intend to add to the original R.A.M.S. Library, selected by strict criteria established by Hans.

The works from the original R.A.M.S. Library are republished by R.A.M.S. Publishing Company in the collection, "The R.A.M.S. Library of Alchemy," with permission of the Estate of Hans W. Nintzel.

If you have a work on Alchemy that you believe should be a part of the R.A.M.S. Library, please contact me through R.A.M.S. Publishing Company.

Philip N. Wheeler

www.ingramcontent.com/pod-product-compliance
Lightning Source LLC
Chambersburg PA
CBHW080812180526
45168CB00006B/2412

* 9 7 8 1 5 1 1 6 6 7 1 6 6 *